URBAN GEOGRAPHY

Urban Geography

AN INTRODUCTORY GUIDE

David Clark

THE JOHNS HOPKINS UNIVERSITY PRESS
BALTIMORE, MARYLAND

First published in Great Britain by
Croom Helm Ltd, 2-10 St John's Road, London SW11

First published in the United States of America by
The Johns Hopkins University Press
Baltimore, Maryland 21218

Library of Congress Catalog Card No. 82-48469
ISBN 0-8018-2965-8 hardcover
ISBN 0-8018-2966-6 paperback

Printed and bound in Great Britain

CONTENTS

FIGURES

Tables

For Judith and Hannah Lucy

PREFACE

There can be few areas of study that have expanded as rapidly in the last eighty years as urban geography. What in 1900 barely existed as a distinctive specialism is today an integral and indispensable component of any undergraduate degree course in geography. Associated with this growth has been a broadening of subject matter and approach such that urban geography now interfaces with most if not all of the disciplines in the social and behavioural sciences. At the same time, a number of varied methodologies and techniques have been developed within, or have been introduced into, urban geography so as to assist in the study of the city. These changes and innovations explain much of the challenge of the discipline. They are also a source of confusion to the student who is confronted by a bewildering array of topics, themes and perspectives. This text reflects a response to the former that will hopefully reduce the latter. It seeks to provide a guide to recent developments and contemporary areas of inquiry in urban geography.

Elements of urban geography are commonly taught in the first or freshman year of undergraduate degree courses in the UK and North America. Towns and cities are examined alongside population, industry, agriculture and transportation within the broad framework of an introduction to human geography. A detailed and specific focus upon urban settlements is normally reserved for the second, sophomore or junior years. Courses at these levels typically seek to outline and review the field of urban geography and to discuss and evaluate in depth, the various theoretical and empirical statements on urban spatial structures. They are commonly followed in the final or senior year by courses which focus upon specific urban topics such as the social geography of the city or urban economics, or which concentrate upon problems, planning or policies towards the city. As an introductory guide, this book is designed to be of use at the first or second year level. The approach is primarily systematic and analytical rather than thematic or applied. Basic concepts in urban geography are introduced, and cities are examined at both intra- and inter-urban scales and from a wide range of alternative perspectives. Only at the end of chapter 7 does the focus become specifically problem-oriented. This brief change of direction demonstrates the value and limitations of academic analysis in understanding the contemporary urban crisis, as well as providing a teaching link between introductory systematic, and advanced thematic approaches, in urban geography.

What appears between these covers is essentially one individual's synthesis and outline of a complex field of study. Little of the substantive material is new, indeed some of it is very old, so that any distinctiveness is a product of selection or organisation. Inevitably, the bias of the author's own interests and experience is reflected in places with the result that some topics are covered in depth while others, such as the Third World city, receive little attention. Rather than overload the text with citations, use is made throughout of tables of selected references. Such lists are not intended to be a comprehensive and definitive guide to the published material, rather, they aim to direct the student towards the most useful and accessible supplementary reading. A second objective is to overcome the locational biases which are an inevitable characteristic of geographical writing. Even though the text may refer to a particular UK or North American city or study, the tables will hopefully identify examples in other countries or areas with which the reader may be more familiar.

This book was initiated and completed during the course of regular employment in the Department of Geography, Coventry (Lanchester) Polytechnic, but was substantially written in a period of study leave in the University of Oklahoma. Thanks are therefore due to two Chairmen, David Smith in Coventry and Neil Salisbury in Norman, for their encouragement and for making the secretarial and technical facilities of their departments available. Colleagues in both institutions gave freely of their assistance and advice and I am much indebted to Hugh Matthews, Michael Healey, Ed Malecki and Dan Fesenmaier for comments on particular points and chapters. John Harlin and John Harrington sustained my non-working hours in Oklahoma with a highly rewarding and most welcome supply of social and sporting distractions. Maps and diagrams were expertly drawn by Mary Merrill. Finally, an apology is in order to several generations of undergraduates in Coventry (Lanchester) Polytechnic, whose courses in urban geography have been characterised by a surfeit of jargon and a deficiency of direction. It is in the hope of guiding their successors more clearly that this book was primarily written.

URBAN GEOGRAPHY

1 THE FIELD OF URBAN GEOGRAPHY

The modern city is the product of an extremely long process of development. Research workers interested in early civilisations have identified a number of settlements as early as the fifth century BC to which they accord the title of city, although these places were invariably small, thinly scattered and easily reverted to village or small town status. Today, over 75 per cent of the population of the UK, USA, Canada and Australia live in centres which are termed 'urban' in the census, and the effects of cities are experienced daily by those who live in the remotest 'rural' areas. Rather than being an exceptional settlement form in a basically rural economy, the city has become the central focus of social and economic activity and influence in modern urban society.

In view of this contemporary importance, it is not surprising to find that research workers have tried to build up an adequate framework of knowledge in order to make it easier to understand and plan for the city. This task has occupied a large number of specialists drawn from a wide range of fields in the social and environmental sciences. No one discipline can claim to monopolise the study of the city since urban problems cut across many of the traditional divisions of academic inquiry. Equally, no single methodology predominates in urban analysis for the complexities of urban life necessitate the adoption of a wide variety of approaches. It is in the interdisciplinary nature of urban problems that the city poses the greatest difficulties to the analyst. Progress in urban understanding requires the fusion of insights derived from a number of disciplines each of which approach the study of the city in their own distinctive way.

Geography is the scientific study of spatial patterns. It seeks to identify and account for the location and distribution of human and physical phenomena on the earth's surface. Emphasis in geography is placed upon the organisation and arrangement of phenomena, and upon the extent to which they vary from place to place. Although it shares a substantive interest in the same phenomena as other social and environmental sciences, the spatial perspective upon phenomena which is adopted in geography is distinctive. No other discipline has location and distribution as its major focus of study. It is the characteristics of space as a dimension, rather than the properties of phenomena which are located in space, that is of central and overriding concern.

1

Geography is scientific in so far as it aims to develop general rather than unique explanations of spatial patterns and distributions. It proceeds from the assumption that there is basic regularity and uniformity in the location and occurrence of phenomena and that this order can be identified and accounted for by geographical analysis. In examining spatial structure, geography focuses upon those distributional characteristics that are common to a wide range of phenomena. To this end, emphasis in geographical study is placed upon models and theories of location rather than upon descriptions of individual features. The overriding aim is to develop an understanding of the general principles which determine the location of human and physical characteristics.

Urban geography is that branch of geography that concentrates upon the location and spatial arrangement of towns and cities. It seeks to add a spatial dimension to our understanding of urban places and urban problems. Urban geographers are concerned to identify and account for the distribution of towns and cities and the spatial similarities and contrasts that exist within and between them. They are concerned with both the contemporary urban pattern and with the ways in which the distribution and internal arrangement of towns and cities have changed over time. Emphasis in urban geography is directed towards the understanding of those social, economic and environmental processes that determine the location, spatial arrangement and evolution of urban places. In this way, geographical analysis both supplements and complements the insights provided by allied disciplines in the social and environmental sciences which recognise the city as a distinctive focus of study.

Types of Urban Geographical Study

The scope and nature of urban geography is best illustrated by reference to the data on towns and cities that typically form the starting point for urban geographical study. Two basic types of data are in fact necessary to summarise the geographical characteristics of urban places, the one relating to land using activities such as population, housing and industry, the other to the different types of exchange, linkage and interaction which take place within and between centres (Figure 1.1). In the formal geographical data matrix, cities or parts of cities are listed down the rows, while variables on the contemporary demographic, economic, political and environmental characteristics of each place are entered across the columns, so that the observations or 'geographical

Figure 1.1: Geographical Data Cubes

Formal Geographical Data Cube

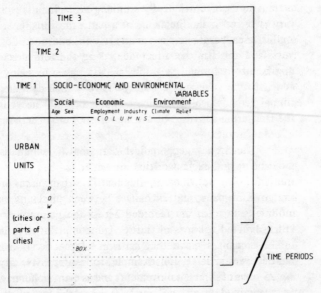

Functional Geographical Data Cube

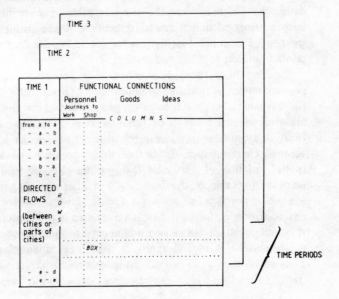

facts' in each cell of the matrix record the extent to which each characteristic is present in each place. It is thus possible by reading along the rows to build up a detailed picture of the social and economic activities present in each centre or, by scanning the columns, to see how places vary in terms of the incidence of a particular variable or characteristic. An historical perspective is provided by the addition of data for previous years and this has the effect of making the formal geographical data matrix into a formal geographical data cube. As many 'slices' can be added to the cube as there are years for which data are available. The formal geographical data cube contains comprehensive information on the land using activities of a wide range of urban activities at a number of points in time.

The functional geographical data matrix summarises the flows, movements and exchanges that are generated by, and in turn maintain, the land using activities. In this matrix, urban places are entered into the rows in pairs, and exchanges between places of personnel, commodities and ideas are recorded across the columns, so that the cells index directed volumes of traffic. The complexities of inter- and intra-urban exchanges mean that matrices of functional geographical data tend to be very large. For example, for a simple five city system, there are 25 rows of directed movements and as many columns as are necessary to accommodate all the traffic flows. Addition of similar data for previous years transforms the matrix into a functional geographical data cube which indexes the characteristics of urban movements in both a temporal and a spatial dimension. These two cubes can be thought of as conceptual devices for organising the data which are of primary interest to urban geographers.

Working from these data cubes, it is possible to identify three primary modes of urban geographical study (Berry, 1964). The first is the 'systematic' approach to urban geography and involves the examination of selected columns of variables so as to see how centres differ from, or are similar to one another in terms of their land use or interactional characteristics. Studies of this type are concerned with the spatial variation of particular features and give rise to distributional maps, as, for example, the distribution of cities according to population size or importance as a traffic focus. Mapping of socio-economic characteristics or flow patterns is a basic step in geographical study and is commonly undertaken as a preliminary to more advanced forms of urban analysis. A second approach is to focus upon selected rows or parts of rows so as to compile an inventory of the socio-economic or traffic mix of individual centres. This perspective is 'regional' in the sense that the city is regarded as an area, and attention is focused upon the totality of socio-economic and interactional variation within it. Examples of this approach are to be found in urban atlases which map a wide range of information about the geographical characteristics of an individual city. The third type of study involves the examination of a 'box' consisting of a number of rows and columns of the matrix. This involves elements of both the 'systematic' and the 'regional' approach and examples are provided by studies which seek to classify urban areas on the basis of their socio-economic or interactional character.

These column, row and box approaches to the data represent the most elementary modes of urban geographical study. More advanced analysis may involve comparative studies in which attention is focused upon the degree of similarity between pairs or sets of columns and rows. For example, the extent to which employment structure relates to the size of cities would be examined by comparing and correlating selected columns of data in the matrix. Similarly, studies of comparative urban structure would focus upon pairs of, or series of rows, so as to see how the socio-economic or interactional character of one place differs from or is analogous to other places. The availability of data for previous years increases the range of approaches considerably since it means that the five types of urban study so far defined can be undertaken in a temporal context. Urban historical geography is, therefore, concerned with identifying and explaining the nature and extent of 'systematic', 'regional' and 'comparative' changes in urban spatial character.

As well as identifying the major types of urban geographical study, the data cubes point to the existence of a number of different levels of geographical inquiry. Data on economic and interactional character may be assembled for many different types of urban unit, including census enumeration districts or tracts, wards, urban places and urban regions, and by changing the entries in the rows, it is possible to study the city at a number of spatial scales. In conceptual terms, the range of scales is infinite and continuous, but geographers tend to focus upon processes and patterns at either the inter-urban or the intra-urban level. In inter-urban studies, towns and cities are the basic units of analysis, and emphasis is placed upon their structural similarities and differences, and the extent to which they interact and interdepend as parts of an urban system. Conversely, at the intra-urban level, the individual city is the focus of attention and analysis is directed towards the identification and explanation of internal patterns of land use and interaction. Some of the most important questions in urban geography surround the extent to which processes and patterns which are observed at one scale are present at the other, and how they are related.

The Changing Focus of Urban Geography

Although the concern with the spatial structure of the city has been central to urban geography since its emergence as a sub-discipline, the various types of urban study outlined above have received different emphasis at different times in the past. Urban geographers have not held a static view of their subject, rather the consensus as to what are the main issues requiring geographical investigation has changed markedly. Such shifts of emphasis are largely a product of changes in the philosophy and methodology of geography as a whole. Whereas geography in the early twentieth century was preoccupied with exploration and discovery, with the relationship between man and his environment and with defining and describing regions, primary attention since 1945 has been placed upon spatial modelling and spatial analysis (James, 1972). This emergence of spatial analysis as an accepted central focus or paradigm represented a fundamental redirection of geographical inquiry which affected all branches of the discipline in the 1950s and 1960s. Today, spatial analysis with its emphasis upon pattern is being increasingly challenged by those who would direct more attention towards the processes that give rise to geographical distributions. Urban geography as a consequence has undergone and is undergoing some major changes of focus from its early concern with the siting and situation of cities through to its present-day interest in the behavioural and political aspects of urban structure (Figure 1.2).

Urban geography has a long tail but a short body. It can claim to have origins in the writings of the ancient Greek scholars, Eratosthenes and Strabo, but its existence as a sub-discipline is very much a feature of the present century, and indeed of the last thirty years. Pre-twentieth century works were strongly descriptive and typically amounted to little more than observations on the physical appearance and subjective impressions of urban places. For example, Martyn's geographical magazine of 1793 was concerned to describe the architectural features of English cities. Similarly, Pinkerton's (1807) 'modern georgraphy', contained a brief account of the chief cities and towns in England according to their dignity, opulence, population and location. These gazetteer approaches were of some value to the traveller, but were not sufficiently systematic or penetrating in their descriptions to form the basis of a scientific urban geography.

Figure 1.2: Major Stages in the Evolution of Urban Geography

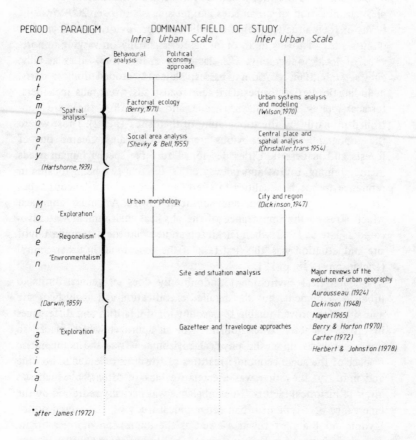

For Carter (1972) it was the replacement of description by interpretation of location which laid the foundations for modern urban geography. From the pioneering works of Hassert (1907) and Blanchard (1911), a form of urban geographical work emerged, which centred around the analysis of the site and situation of towns and cities. This approach was environmental in emphasis and assumed that general location and setting were primary determinants of urban character and that the origins and development of the town were a response to local physical conditions. Site and situation studies typically had two elements; a detailed description of different types of urban site and an analysis of the effects of location and situation upon urban growth and form. The approach is fully illustrated in Taylor's *Urban Geography* (1949) which he claimed to be the first urban text in English. It consisted of a complex classification of urban sites and an analysis of over 200 towns in terms of their situation, relief and climate. For Taylor, it was logical to discuss sites of towns in an order based essentially on varying importance of local topography. His classes were, therefore: cities at hills; cuestas; mountain corridors; passes; plateaux; eroded domes; ports, including fiords, rias, river estuaries and roadsteads; rivers, falls, meanders, terraces, deltas, fans, valleys, islands; and lakes. All of these are 'controlled primarily by the topography of their sites' (p. 12). Those where the topography is less obvious are cities on plains, cleared out of forests and in deserts. Cities developed to serve special human needs include mining, tourist and railway cities, the first two being located in reference to special qualities in the environment. Miscellaneous types are planned, ghost, boom and suburban towns. A similar approach which stressed the importance of the physical environment was advocated as late as 1948, when Dickinson argued that the determination of site and situation was 'the first task of the geographer in urban study' (Dickinson, 1948, p. 12).

The physical environment undoubtedly does set general limits to urban development, but the detailed classification and analysis of site and situation proved unable to account for similarities and differences in urban character. One problem with an approach which placed so much emphasis upon the physical environment was the comparative neglect of the socio-economic features of towns as reflected in housing and industry; the other was the relative lack of attention which was given to historical factors. This imbalance was partially redressed by the emergence of a type of urban geographical study which focused on the layout and build of towns viewed as the expression of their origin, growth and function. Urban morphology studies sought to classify and differentiate towns in terms of their street plan, appearance of building and function or land use. They originated in Europe, especially in Germany, where the rich variety of settlement layouts and building types, which reflect many changes of architectural style, pose especially challenging problems of classification to the analyst. Examples of this approach include Müller's (1931) study of Breslau which proposed a classification of houses based upon roof types, Geisler's (1924) survey of architectural styles in Danzig and Martiny's (1928) essay on the morphology of German towns which distinguished on functional grounds between natural urban forms and those which resulted from conscious urban planning. These early studies in urban morphology were criticised by Dickinson (1948) because the approach was empirical rather than genetic and it is 'only the latter which permits the recognition of the significant' (p. 21). Subsequent work by Conzen (1960, 1962) did much to counter this shortcoming by establishing a basic conceptual framework for urban morphological study, but the approach remained essentially descriptive and proved incapable of leading to general statements and explanations of urban structure and form.

Urban morphology studies were paralleled in the inter-war period by a focus at the inter-urban scale upon the city and its surrounding region. For Dickinson (1948), the urban settlement in general has a two-way relationship with its surroundings that extends beyond its political boundary. First, the countryside calls into being settlements called urban to carry out functions in its service. Secondly, the town, by the very reason of its existence, influences in varying degree its surroundings through the spread of its network of functional connections. Urban-rural links were examined incidentally as part of site and situation approaches, but with the growing realisation of their functional importance, they became a central focus of attention. At one level, city regional studies sought merely to identify urban spheres of influence by mapping indices of commuting (Chabot, 1938), food supply (Dubuc, 1938) and telephone calls (Labasse, 1955), but at another they involved comprehensive investigations of the role of functional regions in spatial organisation. This latter approach was of considerable academic and practical importance and is represented by Dickinson's *City, Region and Regionalism* (1947). City regional studies established a major field of urban geographical inquiry at the inter-urban scale. It was largely in response to questions raised in these studies about urban location and areal servicing that the deductive methodologies were adopted that characterise the contemporary period of urban geography.

The Development of Urban Theory

Despite their very different focus, site and situation, urban morphology and city regional studies placed similar emphasis upon geographical description and inductive inference. They mapped and analysed in detail the environmental, physical and functional characteristics of the town prior to advancing explanations as to how and why particular urban features have arisen. Although it may be possible to say something about general underlying causes, the varied character of cities, and the complex nature of most urban processes, means that this approach left many of the major factors responsible for urban structure and form unexplored. There was, furthermore, a tendency to focus upon the unique circumstance at the expense of the general relationships which apply to all towns and cities. In conceptual terms, descriptive and inductive approaches were superseded in the 1950s by attempts to develop general theoretical statements about geographical distributions using deductive methods, a change in basic methodology which for Davies (1972) amounted to a 'conceptual revolution' in the subject. In contrast to the unsystematic collection of data which is a feature of the early stage of inductive approaches, the deductive route places primary emphasis upon model building (Figure 1.3). Models are idealised representations of reality which demonstrate or summarise some of its properties. They amount to elementary generalisations about the real world which can be extended and refined through testing and reformulation so as to lead to general explanations. Model building approaches were widely adopted in urban geography in the 1960s as Chorley and Haggett (1967) have shown. They were introduced into urban geography through developments in two particular fields; studies of the location, and studies of the internal social and spatial structure, of cities.

Like the analysis of urban morphology, studies of urban location owed much to pioneer work in Germany. Over a century ago, work on the relationship between distance and movement led to elementary models being proposed for agricultural production (Thunen, 1826) and industry (Weber, 1909), but it was in 1933 that Walter Christaller published his great work on *Central Places in Southern Germany* which provided a broad theoretical base for locational analysis in urban geography. The significance of Christaller's work lies not so much in its empirical content, though this is substantial as chapter 5 shows, but in its methodological innovations. Christaller developed a set of interrelated propositions based upon a series of simplifying assumptions and constraints which amounted to a theory of service centre location. This provided a more systematic basis for the study of urban distributions than was possible with either site and situation or city regional approaches. An important characteristic of the theory was that the patterns it predicted were compared with the actual distribution of towns and cities, and as a result, more sophisticated theories of urban location were developed. Central place theory was by no means the only theory of urban location to be developed at this time but it was by far the most important. It demonstrated that fundamental principles and relationships determined the distribution of towns and cities and that these factors could be modelled so as to lead to general theoretical

Figure 1.3: Alternative Approaches to Explantion in Science. Route 1 is the inductive approach, route 2 is the hypothetico-deductive approach.

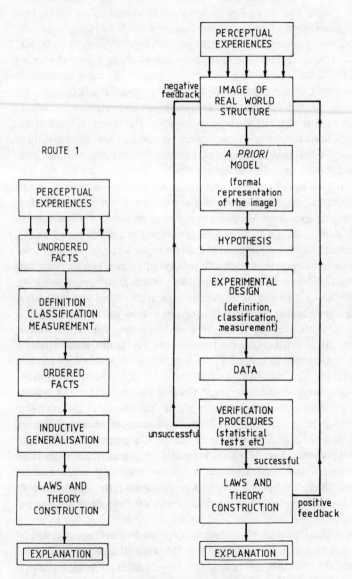

Source: Harvey (1969) p. 34. Reproduced by permission of Edward Arnold (Publishers) Ltd.

explanations of urban location.

That central place theory was capable of mathematical expression meant that much of the work that followed from it was strongly quantitative in nature. At one level, quantitative techniques are used by urban geographers to measure the goodness of fit between predicted and observed characteristics, and to add objectivity to such traditional tasks as city classification and urban regionalisation. At another, they provide a powerful language by which urban systems can be analysed and modelled. Systems modelling in urban geography seeks to identify the structure and state of an assemblage of towns and cities which interact and interdepend as an operating unit. The central feature of the approach is its emphasis on the system of cities as a whole rather than upon any individual or group of centres. Despite the obvious complexity of modern urban systems, which can assume a wide range of states, it can be shown that one particular configuration is likely to overwhelm all the others. Once this configuration has been identified, verified and expressed mathematically, the model can be run using growth data as inputs so as to predict patterns of future change. In this way, urban geography can provide a direct input into the planning field.

Of central importance to the development of this area of study was Wilson's (1970) work on the most likely state of urban systems. An example is the problem of establishing the traffic flow within a system in which the number of trips originating in a series of residential areas is known, as is the number ending in each of a series of work place areas, but the pattern of movements between the two is unknown. Drawing heavily upon work in statistical mechanics, Wilson developed a method of predicting the most likely overall state of the system by maximising the uncertainty, or entropy, about the micro-states within the system. Entropy maximising techniques are highly mathematical and are by no means easy to understand, but Wilson's basic approach was acclaimed by Gould (1972, p. 689) as 'one of the most imaginative and thought provoking works in geographical literature'. It stimulated a wide range of extensions and applications in urban and regional modelling as for example those of Batty (1976), and established the importance of urban systems modelling as a central focus of study in modern urban geography.

These developments at the urban systems level were paralleled by changes in the nature of urban study at the intra-urban scale. Analysis of the internal structure of towns and cities had traditionally been restricted to morphological study, but a major innovation of the early 1960s was a revival of interest in the work of the Chicago School of

Human Ecology, and a more widespread realisation 'that *people* lived in towns and cities and that there was order in residential patterns' (Herbert and Johnston, 1978, p. 14). Elementary models of urban land use structure had been proposed some years previously by Burgess (1925) and Hoyt (1939), but these were superseded in 1955 by Shevky and Bell's social area analysis which presented a general theory of urban social change and predicted the implications for the social and economic structure of the city. Social area theory was far more loosely structured than central place theory but it had a similar catalytic effect. It inspired a wealth of studies which were concerned either to confirm or extend the analysis, or to reject it and replace it with an alternative theory. At first, tests of the theory were confined to simple mapping and elementary statistical analysis of small scale census data, but with the increasing availability of high speed computers, geographers in the last decade have been able to undertake more advanced quantitative studies. So much use was in fact made of the technique of factor analysis that for a time it became almost a standard procedure for identifying the social and spatial structure of cities. Factorial ecology represented an important technical advance in urban social geography, but the same basic focus, implicit in the spatial analysis paradigm, was maintained. The aim remained one of testing and reformulating theoretical explanations that seek to account for the social and spatial structure of the city.

Contemporary Urban Geography

Where stands urban geography today? In a recent review, Herbert and Johnston (1978) identify two contemporary influences which, if they continue to gain in momentum, could form the basis of new paradigms within the discipline. The first is the behavioural approach, and involves a concern with the ways in which individuals perceive and make decisions about the city. The second is the political economy approach which seeks to explain urban processes and account for urban problems by reference to alternative political ideologies. These emerging strands represent, in their very different ways, reactions to two aspects of the spatial analysis approach; the overemphasis upon normative behaviour and the analysis of urban patterns, and the neglect of political values and relationships in most urban and regional models.

Behavioural perspectives seek to redirect urban geography towards the processes which give rise to the urban pattern. The rationale for this approach is that urban structure cannot be fully understood by simply finding two similar spatial patterns (as between income and housing). Such explanations are inadequate because they correlate two phenomena without specifying the link through which they are associated. They do not answer the question of how and why such associations exist. Behaviourists 'believe that the physical elements of existing and past spatial systems represent manifestations of decision-making behaviour on the landscape and they search for geographic understanding by examining the processes that produce spatial phenomena rather than by examining the phenomena themselves' (Amedeo and Golledge, 1975, p. 348). They recognise that activities and land use patterns do not just happen, but arise from decisions made by individuals and organisations, both private and public.

In analysing urban processes it is acknowledged that decisions are mostly taken in conditions that are far from ideal. Individuals rarely have complete and comprehensive information about the circumstances which affect their daily lives, so that most judgements which they arrive at are sub-optimal. This is especially true in the case of city dwellers who must organise their activities within a complex and varied urban environment. Although cities exist as physical objects, as collections of buildings, highways and open spaces, there is no real evidence that they are perceived by the people living in them in the same way in which they are objectively structured. For the behaviourists, the city has both a physical structure and a psychological or cognitive structure, and it is the latter which determines how individuals will respond to, and interact with, urban places and institutions. In contrast to location models of the central place variety, which are based upon crude assumptions about human behaviour at the aggregate level, behavioural approaches explore the ways in which people structure their images so as to create mental maps of the city. They further trace the influence of these perceptions for major areas of decision making such as shopping, housing choice and migration.

The study of perception and cognition has a long tradition in psychology, but was first introduced into urban geography through Lynch's (1960) analysis of residents' images of Boston, Massachusetts. Subsequent work includes that of Downs and Stea (1973, 1977) and Gould and White (1974). Decision making studies have a similar history and trace their origins to Wolpert's (1964) study of agricultural productivity, Pred's (1967, 1969) behavioural models of agricultural, retail and industrial location and Rushton's (1969) analysis of consumer behaviour. Despite the wide range of ongoing research, the comparative recency of these developments means that the contemporary status of behavioural approaches in urban geography is unclear. Cox and Golledge's (1969) claim that a behavioural revolution occurred in the late 1960s was almost certainly premature. Rather than represent a new paradigm, the full impact of behavioural approaches upon urban geography has most probably yet to be realised.

The second contemporary influence has origins in political economy. It arose out of a growing concern for the acute and persistent social and economic problems which are a feature of many cities and especially inner areas of large cities in 'advanced' Western economies. Geographers in the 1970s were extremely active in measuring and mapping poverty, homelessness, deviance and unemployment, but although spatial analysis could identify some basic ecological relationships, it proved unable to account for urban deprivation and, more important, to suggest appropriate remedies. One reason is that many analytic models are based upon assumptions of private ownership, private enterprise and profit making which many argue are the root causes of social inequality and conflict in the city. The other is that the role of government departments, planning agencies and financial institutions, which exercise a powerful and possibly a biased control upon urban economy and society, is overlooked.

Two areas of specific interest can be identified within the broad context of contemporary political economy approaches. The first is essentially concerned with the analysis of conflict and management in the city, and the ways in which individuals and interest groups define and pursue their geographical objectives. It examines the nature of inter-group attraction and aversion, and the role of urban 'gatekeepers' (such as bank managers, building society managers, housing officers and estate agents) who can use their financial and political power to determine, or at least to control, the structure of the city. The second approach is altogether more radical, and derives its major inspiration from the writings of Karl Marx. It seeks to introduce an alternative philosophical stance and research direction to urban geography. In *Social Justice in the City,* Harvey (1973) argued that social injustices are an inherent feature of societies organised on a capitalist basis, and that changes in urban structure can only be achieved through the removal of private ownership and, in particular, monopoly rights over land. Peet (1978) has gone further by calling for geographers to seek alternative environmental designs and to replace bureaucracies by anarchistic models of community control. The implications for urban geography are profound. No longer can it be a neutral academic discipline dedicated to explaining the spatial structure of the city; instead it must become an active and committed focus for urban social and geographical change.

Conclusion

This overview has identified the major perspectives which have been adopted by urban geographers in recent years. The picture which emerges is one of a dynamic and rapidly changing discipline which has experimented freely with a wide range of approaches to urban geographical explanation. Such vigour is in many ways a healthy and encouraging sign, but it is also a source of considerable confusion and difficulty to the student. One problem is the seemingly continual need to master new and unfamiliar methodologies and techniques so as to follow the latest developments in the subject. The other is the task of linking the various perspectives together so as to form a coherent understanding of the ways in which urban geographers study the city. The situation is not helped by the recent proliferation of undergraduate texts which, while ostensibly concerned with urban geography as a whole, seek in reality to present a particular viewpoint, be it spatial analysis, behavioural or political science, at the expense of the others.

This guide seeks to introduce urban geography as a field of study. No one perspective is emphasised, the aim is rather to outline and assess the contribution of urban geographers to the understanding of the city and of urban society. Chapter 2 examines the basic problem of defining towns and cities, placing particular emphasis upon the different ways in which urban settlements are viewed by those who live in them, and those who plan, administer and study them. Chapter 3 is concerned with urban growth especially in the industrial and post-industrial periods. Chapter 4 looks at various theories of urbanisation through a detailed investigation of the impact of cities upon society. The factors which determine urban structure at the intra- and inter-urban scales are discussed in chapters 5 and 6 respectively. Finally, urban problems and urban planning are examined in chapter 7. The breadth and depth of urban geography, however, means that the material introduced is highly selective. In place of comprehension, this introductory guide can at best help to identify the major routeways through the discipline. If, like good travel literature, it creates an impression that there is much of interest over the nearest hill or around the next corner, it will have achieved its purpose.

2 DEFINING URBAN PLACES

Towns and cities are the objects of urban geographical inquiry. They are of interest in so far as they vary from place to place, are differentiated internally and evolve over time. To the geographer, the city is a unit of analysis consisting of a collection of buildings, activities and population clustered together in space. It can be distinguished from other forms of settlement in terms of the density of concentration of these attributes. Indeed, it is by specifying and mapping such indices that the urban patterns can be identified and urban processes studied. Urban residents, however, rarely take such a detached view of cities in which they live. For them, the city is a collection of symbols and values based upon familiarity, impression and personal experience. An understanding of urban cognition is important since it exercises a major control upon urban spatial behaviour. An individual's image of the city determines whether it is liked or abhorred, and where within it, if at all, he will choose to live, shop, work and play. The basis of both subjective impressions and objective definitions of the city are considered in this chapter.

Perceptions of the City

When viewed from the air, most large cities have a clearly discernible form. Physical geography characteristically imposes a basic pattern by way of the distribution and configuration of relief features, rivers, coasts and shorelines. Against this backcloth, a cluster of high buildings identifies the central business district, widths of roads and the alignment of canals and railways demarcate the major arterial routeways, while concentrations of factories, shops and housing point to the existence of industrial, retailing and residential areas. The density of building is usually lowest around the periphery so that even though it may not be possible to specify a precise dividing line, a narrow zone of transition between urban and rural land use is normally apparent. Although the process of perception is the same, the perspective from ground level is very different. Lacking the benefit of a grand overview, the city takes on a far more intimate and personalised identity. Individuals receive visual cues from the urban environments with which they are familiar,

17

and arrange these perceptual signals into mental maps which, for them, define the city. Which elements of the urban landscape most commonly form the basis of individuals' perceptions? It was this question that was first addressed by Lynch in his book, *The Image of the City* (1960).

This work was concerned with the visual quality of three American cities as revealed by an examination of the way the physical features of their landscape were perceived by their residents. The central concept was the 'legibility' of the urban environment, or the ease with which individuals can organise the various elements of urban form into coherent mental representations. Lynch hypothesised that cities varied in the extent to which they evoked a strong image – a quality he termed 'imageability' – but it was most likely that cities that were 'imageable' were places that could be seen as patterns of high continuity with interconnected parts. In other words, a city was likely to be 'imageable' if it was also 'legible'.

This hypothesis was tested in Boston, Los Angeles and Jersey City. In each place, respondents were asked to draw a quick map of the central area 'just as if you were making a rapid description of the city to a stranger, covering all the main features'. They were also asked to list the elements of the city they thought most distinctive and to locate, describe and express any emotional feeling towards any part of the urban landscape. The findings for individual respondents were aggregated and then compared with visual surveys which had been carried out by trained observers. On this basis, Lynch classified the contents of a city's image into five types of elements (Figure 2.1).

Paths are the channels along which we customarily, occasionally or potentially move within the city. They range from streets to canals and are reference lines we use to arrange other elements. *Edges* are line breaks in the continuity of the city. The most pronounced edges are usually prominent, continuous in form and impenetrable. An edge may take the form of a river, a railway line, a large wall or even a forested green space. *Nodes* are focal points within the city. They are commonly road junctions or meeting places where activities are concentrated. Highly visible structures surrounding a node enhance its positive image. A shopping area or an office building is an illustration of a node. *Districts* are medium to large sections of a city which are identified by some common character and which the individual can enter. Beacon Hill and South End in Boston are typical districts. Finally, *landmarks* are also reference points but are much smaller in size than nodes. A landmark is usually a single physical object such as a building, a store or a mountain. It may be notable for its beauty or ugliness. Big Ben, London and

Figure 2.1: Lynch's Five Major Elements of Mental Maps.

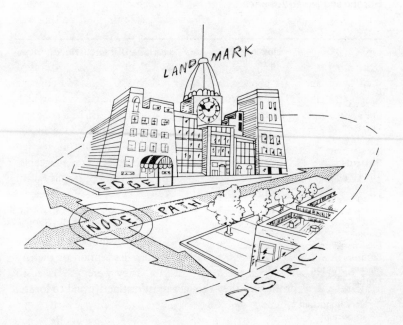

Source: Porteous (1977), p. 103.

the Eiffel Tower, Paris illustrate the role of landmarks. Studies using Lynch's approach have confirmed the importance of these elements in individuals' perceptions of the city (Table 2.1).

These elements are structured together as mental images in complex ways. Clearly, none of the element types exist in isolation. They overlap, interdigitate and form clusters or complexes. A market square for example is not only a visually distinctive area in the city but is also seen as a node, the meeting place of paths, a set of edges and the location of landmarks. The range and character of elements present is clearly itself a factor in image building. For example Haggett (1975, p. 532) has argued 'that we construct mental maps more readily in older European cities (a baffling jumble of streets but with distinctive elements to serve as locational clues as to where we are) than in a new North American city (a regular street pattern but too many roads looking like each other)'. Cities with strongly developed elements may be seen as interesting

Table 2.1: Importance of Landscape Elements in City Image Studies in Europe and North America

City	Sample Size	Paths	Landmarks	Districts	Nodes	Edges
Boston[1]	30	+	*	*	+	+
Jersey City[1]	15	+				
Los Angeles[1]	15	*	+		+	
Chicago[2]	42	*	*	*		
Englewood[1]	44	*	*	*	+	
Amsterdam[4]	25	*			*	
Rotterdam[4]	22	*	+		+	
The Hague[4]	25	*	+	+		
Rome[5]	47	*	*	+	*	
Milan[5]	41	*	*	*	*	
Birmingham[6]	167	*	*		*	
Hull[7]	95	*	*		*	
Goole[8]	20	*	*	*	+	
Ellesmere Port[8]	25	+	+	*		
Stourport[8]	20	*	+	*	+	
Market Drayton[9]	43	*	*			
Durham[10]	94	*	*	+		

* = very important
+ = important

All relate to mapping exercises

Notes: 1 Lynch (1960); 2 Saarinen (1969); 3 Harrison and Howard (1972); 4 de Jonge (1962); 5 Francescato and Mebane (1973); 6 Goodey *et al.* (1971); 7 Goodey and Lee (1971); 8 Porteous (1971); 9 Goodchild (1974); 10 Pocock (1975).

Source: Pocock and Hudson (1978), p. 51.

environments in which to live, while those with weak elements may be formless, monotonous, lacking in character and, therefore, unattractive.

What Lynch emphasised was that individual images combine and overlap to provide a public image of the city. It is likely that individuals belonging to a fairly homogeneous group in a particular area will have mental maps which are not entirely unrelated. As a consequence, each unique map will interdigitate with others in varying ways so that an overall public image becomes apparent. Experience supports this view. Trafalgar Square, London is readily distinguished because of its mix of distinctive buildings, statues and monuments, and although seen slightly differently by each person, is likely to figure sufficiently in each individual's image as to be a prominent feature in the collective perception of the city. Public images of Boston, as derived from verbal interviews and sketch maps, are shown in Figure 2.2 where the strength of the symbol is related to the frequency of mention by Lynch's interviewees.

Important differences characterise the city as perceived and the city as objectively defined (as for example by a street map). Public images are warped in ways which emphasise the familiarity of home territory, the central area, and routes between the two. Major landmarks tend to dominate the collective image and central areas are perceived especially clearly, probably because it is this part of the city that is visited most frequently by the urban population as a whole. These conspicuous urban features provide a basic means of direction finding in the city so that in most cases, people orient themselves in relation either to their homes or to the city centre rather than to the cardinal points of the compass. Another prominent feature of public images is their tendency to 'better' the environment, to record a structure more uniform and less haphazard than that present in reality. Such enhancement is not restricted to those who are learning about the city but it seems to be a general characteristic. This suggests an important practical application since studies of perception may be used to identify areas in the city with a poor image and to suggest ways in which their appearance could be altered or improved as, for example, to attract population, industry or tourists.

Lynch's methodology has been criticised on three main grounds. The first is that it presupposes that individuals hold a map, that is, a spatial representation of the city in their minds, whereas urban images may in fact be structured with respect to colour, textures, aesthetics, quality or visual appeal, all of which are essentially non-geographical in character. Support for the mental map basis of urban imagery is, however, provided by the work of Appleyard (1970) who found that the maps of 300 Venezuelan respondents were either dominated by sequential elements (notably roads), or spatial elements (shapes, areas and links). The former ranged in complexity from the most fragmentary, which consisted of broken paths and lists of unconnected elements, through maps which portrayed chains and branched features, to elaborate representations which featured hierarchical networks of elements. Similarily, whereas the most simple 'spatial' maps included primitive scatter patterns, the most sophisticated involved complex features which were drawn with precise shapes and boundaries. Although the terminology is different, this supports Lynch's argument that images of the city are essentially geographical and are composed of a mix of simple components. That

Figure 2.2: Images of Boston as Derived from (a) Verbal Interviews and (b) Sketch Maps.

(a)

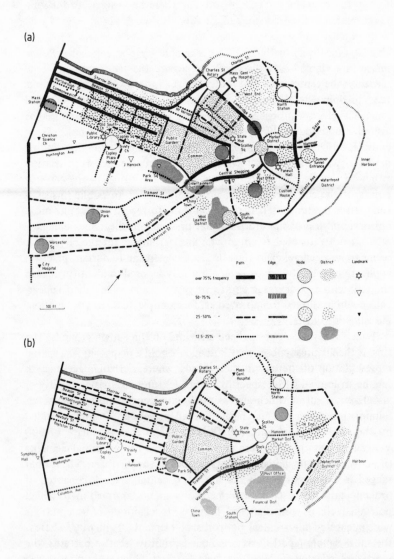

(b)

Source: Lynch (1960), p. 146. Reproduced by permission of Macmillan Press, London and Basingstoke.

33 per cent of Appleyard's respondents based their maps upon chain structures, and 21 per cent upon branch and loop patterns, emphasises further the importance of paths and routeways in city imagery.

The second criticism, applicable to both Lynch's and Appleyard's work is the reliability of the 'sketch map' approach given the highly variable levels of basic graphicacy within the population. Image elicitation requires individuals to structure their perceptions in the form of a mental map and then to outline them on paper, and there may be many people who form clear images of the city within which they live but who are simply unable to draw them. There can in addition be no assurance that a map drawn for a stranger would necessarily correspond with an individual's personal image: it is likely, for example, to place more emphasis upon transport links. The final problem is that of aggregating individual responses and generalising from them, which is especially difficult in mental map studies in view of the highly variable and personal nature of each sketch. It further introduces a second perceptual component — that of the researcher — into the study. These difficulties of construct formation, elicitation and generalisation are common to much psychological research. They emphasise the problems which arise when attempting to undertake objective work in the area, and point to the need for collaborative and follow-up studies.

Despite this interest in aggregate patterns, and their implications for planning, much recent research, reviewed by Porteous (1977) and Pocock and Hudson (1978) has been directed towards evaluating the factors which relate to, and determine, individual differences in the perception of the city (Table 2.2). Sex appears to be an important variable, as a number of studies have revealed significant contrasts in the ways in which males and females construct and retain images. For example, Appleyard (1970) found that females tended to draw area-based rather than path-based (sequential) maps and were more prone to error. Similarily, in a Los Angeles study (Orleans and Schmidt, 1972), wives' maps of the local neighbourhood were found to cover nearly twice as much area as those of their husbands. Such differences may be physiologically based or they may be culturally determined: as males and females participate in different patterns of daily activity, they visit, and so become familiar with, different parts of the city. Experience is of course a critical consideration and means that visitors and recent immigrants are likely to have images of the city that are different to those of long-term residents. The effects of age are of particular interest as cognitive ability is very much a learned skill. There is evidence, for example, that while few six year olds have much sense of place, by the

Table 2.2: Factors Which Affect Individuals' Perceptions of the City: Selected Studies

Factors	Location of Study	Study
Sex	Los Angeles	Everitt and Cadwallader (1972) Orleans and Schmidt (1972)
Age	England Coventry Netherlands Worcester, Mass.; Rio Piedras, Puerto Rico	Porteous (1977) Matthews (1980) Dijkink and Elbers (1981) Blaut *et al.* (1970)
Experience	New York Cuidad Guyana, Venezuela London	Milgram *et al.* (1972) Appleyard (1970) Canter and Tagg (1975)
Socio-economic class	Market Drayton, England Milan, Rome	Goodchild (1974) Francescato and Mebane (1973)
Ethnicity	Harrisburg, Texas Boston	Maurer and Baxter (1972) Ladd (1970)

Sources: Porteous (1977); Pocock and Hudson (1978); Gold (1980).

age of nine the capacity to create spatial images is being acquired (Dale, 1971). Similarly, Matthews' (1980) study of children's perceptions of central Coventry shows that the development of perceptual powers continues through adolescence into early adulthood. Thus for 11 and 12 year olds, Coventry's shopping precinct largely delimits the city centre whereas 17-18 year olds incorporate both a wider area and a greater range of elements in their perceptions. Members of different ethnic groups also possess contrasting images of the city as a result of past traditions and cultural influences (Ladd, 1970). Finally, socio-economic class is a factor. Both Goodchild (1974) and Michaelson (1970) emphasise the fact that the middle class are especially interested in the aesthetic and historical detail of cities. In Rome and Milan, group differences between middle and lower class residents have been identified (Francescato and Mebane, 1973). The middle class tend to place more emphasis upon districts whereas the reverse is true of lower class groups.

That perception varies with sex, age, length of residence, social class and ethnicity, however, does not mean that the precise effects of any of these variables is understood. Indeed there remains much confusion in the literature. The city is in one sense a subjective image in the human imagination. Geographers have yet to determine how such images are formed and in what way urban perception determines behaviour.

Independently of any subjective impression, the term 'city' also refers to a particular type of settlement which is capable of precise statistical and geographical specification. While this task of definition is not one that would unduly trouble the layman, it is of vital academic and practical importance. Each year, governments around the world compile and release a wide range of information about towns and cities, and it is with reference to these data that urban concepts are formulated and tested by research workers, and policy decisions are taken by politicians and planners. There is little point in comparing urban statistics in, say, the United States and the United Kingdom, if the census in these two countries defines towns and cities in completely different ways. Similarly, studies of urban growth are dependent for their reliability upon the accuracy of historical compilations of urban data. Some consideration of the statistical and geographical meaning of cities is necessary in any introductory guide to urban geography.

Objective Definitions

Towns and cities are inherently difficult to define because they are members of a continuum of nucleated settlements that grade into one another. Basic distinctions can be drawn between hamlets and towns, villages and cities, and yet differences are not so sharp between adjacent members of the continuum. Few would argue with the fact that cities have a larger size and greater degree of functional specialisation than towns, but it is no easy task to separate the two precisely. Most would agree that settlements with populations over 100,000 are probably cities, but the status of places with around 20,000 population is more questionable, especially when such places have important local government and commercial functions. The same uncertainties and differences of opinion surround the distinction between hamlets and villages, metropolises and megalopolises. As one goes down the scale from the largest urban agglomeration to the smallest farmstead, it is extremely difficult to identify dividing lines and terminology that are universally acceptable.

The problem of definition arises because towns and cities may differ from rural centres in so many ways. Urban settlements are not only bigger and more densely populated than villages, but they commonly have a wider range of industrial, administrative and government functions and responsibilities as well. They may even enjoy the legal title of 'city', and all the ceremonial trappings that go with it, as a result of an historic act or charter, although such a designation may be of no practical importance today. At first glance, population size would seem to be the most suitable indicator of urban status, but this measure is used in only 33 of the 133 countries and sovereign territories from which data are assembled by the United Nations (Table 2.3). Moreover, the different population minima employed, which range from 200 in the case of Denmark, to 10,000 in Switzerland and Senegal, raises questions concerning the utility of a size measure alone. In 20 cases, population size is combined with other diagnostic criteria. Examples include Zambia and Zaire which use population size and employment in non-agricultural occupations, the United States and India, which employ size, density and administrative criteria, and Bangladesh where consideration is given to both the population and the physical characteristics of a place. Such composite indices provide a more satisfactory basis for urban definition within the country concerned, although the composition and threshold values used again mean that cross-national comparison of urban patterns and problems may be difficult.

The vast majority of towns and cities throughout the world are identified on a legal, administrative or governmental basis. No attempt is made to establish objective criteria, rather urban places are designated by municipal or corporate status. This practice is especially common in the smaller countries of Africa and Central America where towns and cities are designated by government decree, but it applies in several larger and more advanced nations as well. Thus Gaberone, Labatsi and Francistown are cities in Botswana, while Tripoli, Benghazi, Beida and Derna are the only appointed urban places in Libya. Similarly, urban places in Brazil, Thailand and Iraq are designated on a legal-administrative basis. The compilers of the United Nations *Demographic Yearbook* have been severely critical of such definitions. They argue that as the boundaries that mark off towns, cities and municipalities for administrative and local government purposes vary from place to place, are in many cases the result of historical accident, and are rarely altered to comply with population change, they constitute an extremely poor basis for urban definition (UN, 1955). Faced with these difficulties, research workers concerned with urban problems at the international scale have no alternative but to make up their own definitions of towns and cities and try to apply them uniformly across the world. This was the approach

Table 2.3: Selected Definitions of Urban Places Throughout the World.

Criteria	Number	Examples
Population size	33	Denmark: agglomerations of 200 or more inhabitants. Switzerland: communes of 10,000 or more inhabitants, including suburbs. Austria: communes of more than 5,000 inhabitants. Senegal: agglomerations of 10,000 or more inhabitants. Venezuela: centres with a population of 2,500 or more inhabitants.
Population size plus additional criteria	20	United States: places of 2,500 inhabitants or more incorporated as cities, boroughs (except in Alaska), villages and towns (except towns in New England, New York and Wisconsin) but excluding persons living in rural portions of extended cities; the densely settled urban fringe whether incorporated or unincorporated of urbanised areas, unincorporated places of 2,500 inhabitants or more. India: towns (places with municipal corporation, municipal area committee, town committee, notified area committee or cantonment board); also, all places having 5,000 or more inhabitants, a density of not less than 1,000 persons per square mile or 390 per square kilometre, pronounced urban characteristics and at least three-fourths of the adult male population employed in pursuits other than agriculture. Zambia: localities having 5,000 or more inhabitants, the majority of whom will depend on non-agricultural activities. Sweden: built-up areas with at least 200 inhabitants and usually not more than 200 metres between houses. Bangladesh: centres with a population of 5,000 or more inhabitants with such urban characteristics as streets, plazas, water supply systems, sewerage systems, electric light, etc. Zaire: agglomerations of 2,000 or more inhabitants where the predominant economic activity is of the non-agricultural type and also mixed agglomerations which are considered urban because of their type of economic activity but are actually rural in size.
Legal, administration and governmental	65	Egypt: governorates of Cairo, Alexandria, Port Said, Ismailia, Suez; frontier governorates and capitals of other governorates as well as district capitals.

Table 2.3 continued

Gambia: Banjul only.

Morocco: 117 urban centres.

Brazil: urban and suburban zones of administrative centres of municipos and districts.

Iraq: the area within the boundaries of municipality councils.

Hungary: Budapest and all legally designated
towns.

Thailand: municipalities.

United Kingdom:

England and Wales: areas classified as urban
for local government purposes, i.e. county
boroughs, municipal districts and urban
districts.

Northern Ireland: administrative county
boroughs, municipal boroughs and urban
districts.

Scotland: cities and all burghs.

Botswana: the cities of Gaberone and Labatsi
and the urban agglomeration of Francistown.

Libyan Arab Jamahiriya: total population of
Tripoli and Benghazi plus the urban parts of
Beida and Derna.

Not available	15	Ivory Coast. Guyana.
Total	133	

Source: UN (1977), Table 6, pp. 182-6.

used by International Population and Urban Research of Berkeley, California. They defined a metropolitan area as a concentration of at least 100,000 people consisting of a principal city together with any adjacent and dependent commuting area in which at least 65 per cent of the economically active population worked in non-agricultural activities. Despite the poor quality of many national censuses, the team defined some 1,064 metropolitan areas in the world in 1950 and managed to assemble reasonably adequate statistical data for just over four-fifths of them (Gibbs and Schnore, 1960).

Administrative Definitions

Some of the characteristics of administrative definitions of urban places, and the problems they give rise to, can be seen by reference to United Kingdom censuses. Prior to the reform of local government in the

United Kingdom in the 1970s, urban areas were defined as administrative areas described as urban, that is a county, municipal, metropolitan or London borough, a Scottish burgh or an urban district. The remaining administrative areas, called rural districts in England, Wales and Northern Ireland, and districts of counties in Scotland were regarded as being rural. On the basis of this definition, the urban population of Warwickshire in 1971 consisted of all those people living in Coventry, Birmingham and Solihull County Boroughs, Royal Leamington Spa, Warwick, Stratford-upon-Avon, Sutton Coldfield, Nuneaton and Rugby Municipal Boroughs and Kenilworth and Bedworth Urban Districts. The rural population was the inhabitants of Alcester, Stratford-upon-Avon, Shipston-on-Stour, Southam, Rugby and Atherstone Rural Districts (Figure 2.3). With the exceptions of the New Towns, and the officially designated Scottish 'cities' of Edinburgh, Glasgow, Aberdeen and Dundee, the terms 'town' and 'city' were not used for census purposes in 1971. Towns and cities could not therefore be distinguished administratively or geographically. Similarly, there was no way of identifying individual villages or hamlets. In addition to these urban administrative units, seven conurbations – Greater London, South East Lancashire, West Midlands, West Yorkshire, Merseyside, Tyneside and Central Clydeside – were recognised in 1971. These were first defined in the 1951 census as aggregations of entire local authority areas in those parts of the country where there was contiguous urban development.

During the 1970s the structure of local government in the United Kingdom, outside London, was changed. As a result, a new basis for urban definition was introduced although data based upon the pre-1974 areas have been published in the 1981 census. The new administrative areas consist of administrative counties which are divided into districts. No distinction is made in this system between urban and rural areas, and as the new districts are much larger than those they replaced, they commonly contain both towns and cities, and extensive tracts of rural area. Thus in central Warwickshire, Warwick, Royal Leamington Spa and Kenilworth are now included together in Warwick District (Figure 2.3). Similarly, Rugby District includes both the town of Rugby (population 52,000) and a large number of surrounding villages. A different division of administrative responsibilities applies in the six metropolitan counties, but the boundaries of these conurbations are no more realistically drawn. Indeed, as the example of Coventry (population 330,000) shows, the effect of its inclusion within the West Midlands Metropolitan County is to

Figure 2.3: Local Authority Areas in Warwickshire and West Midlands.

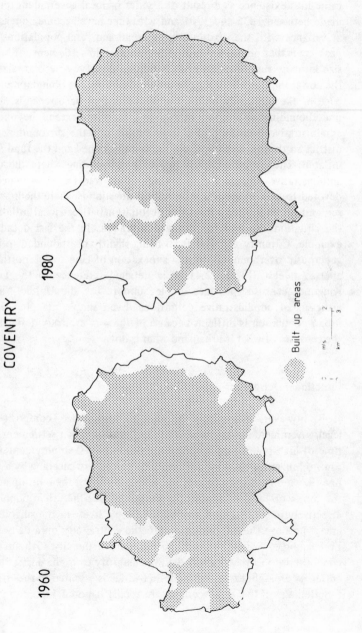

Figure 2.4: The Boundaries of Coventry in Relation to the Built-up Area, 1960 and 1980.

cut it off completely in administrative terms from its extensive commuting field in Warwickshire.

The administrative divisions upon which these definitions are based
came into existence as a result of a series of local government reforms
made between 1835 and 1901 and with time have become increasingly
at variance with the distribution of urban and rural populations. One
problem is that population growth has resulted in settlements of 'urban'
size in many rural districts and districts of county. A second relates to
the correspondence between the urban administrative boundary and the
edge of the physically built-up area. A frequent occurrence is that of
underbounding where the houses and factories extend beyond the
administrative boundary of a large centre into the surrounding rural
district so that the urban population is undervalued and the rural figure
inflated. The opposite case is that of overbounding where the urban
administrative boundary encloses significant tracts of land which are
devoted to non-urban uses. The critical question is where the limits are
set, and as this is often a matter of negotiation between local authorities,
the situation can and has changed overnight. In the last decade for
example, Coventry changed from being slightly overbounded to being
generously overbounded, by the annexation of land to the north-west
and south-east (Figure 2.4). This detailed reference to the United
Kingdom censuses provides ample support for the United Nations
criticism of administrative definitions. Even simple mapping of the
urban population is difficult because of the very arbitrary definition in
the censuses of what is urban and what is not.

Functional Definitions

It is, however, increasingly difficult to justify size, composite and
legal/governmental/administrative definitions of urban settlements at a
time of high population mobility and decentralised employment. Cities
can no longer be circumscribed with any degree of meaning by a single
line drawn on a map to correspond with the edge of the built-up area or
the limits of local government responsibility, as the urban population
effectively includes a large number of people living in the surrounding
area who nevertheless commute into the central city on a daily basis.
For example, 42,000 workers, 30 per cent of the city's labour force,
commute into Coventry every day, the majority from the adjacent local
authority areas; but this massive influx which is essential to the efficient
functioning of the city economy is totally ignored in conventional

physical and administrative definitions of the city. In place of the straightforward physical city, the functional urban region or the urban labour market seems to be a more meaningful unit of analysis for urban geographers.

As long ago as 1920, the United States census recognised the inadequacies of a simple twofold distinction between urban and rural and introduced a third category, that of the rural non-farm, to take account of the growing population living in rural areas but working in cities. This functional definition of urban communities was expanded in the 1940 United States census into the Standard Metropolitan Region, subsequently referred to as the Standard Metropolitan Area in 1950 and the Standard Metropolitan Statistical Area (SMSA) in 1960. The American definition recognises that in functional terms, an urban area consists of a core municipal area linked by journey to work movements to a commuting hinterland. The basic building block of the SMSA is the administrative county which is generally much smaller in population than the United Kingdom county. 'Central cities' or urban cores are defined as counties with at least 50,000 population. Linked to the central city is a 'metropolitan ring' composed of contiguous minor civil divisions with at least 75 per cent of their economically active population in non-agricultural occupations, 15 per cent commuting to the core and finally, a population density of at least 150 persons per square mile. On this basis, 243 SMSAs have been defined in the United States. The utility of these areas is shown by the fact that most statistical surveys of urban change in the United States now use the SMSA as their basic unit of analysis.

In recognition of the value of such functional approaches, geographers have sought to define urban United Kingdom in terms of a similar set of daily urban systems consisting of employment centres and journey to work movements. A modified version of the SMSA concept was first applied to England and Wales by Political and Economic Planning using 1961 census data (Hall *et al.*, 1973), and was extended and updated to cover the whole of the United Kingdom by Drewett *et al.*, (1975). The system is based upon two building blocks defined in terms of local authority administratives areas:

(1) The Standard Metropolitan Labour Area (SMLA) which has a minimum overall population of 70,000 and comprises:
   (a) an SMLA core consisting of an administrative area or number of contiguous areas with a density of five jobs or over per area; or a single administrative area with 20,000

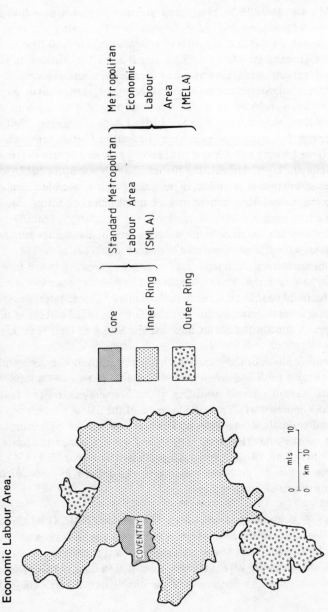

Figure 2.5: Functional Definitions of Urban Areas: The Coventry Standard Metropolitan Labour Area and Metropolitan Economic Labour Area.

or more workers;

(b) an SMLA ring consisting of administrative areas contiguous to the core and sending 15 per cent of their economically active population to that core.

(2) The Metropolitan Economic Labour Area (MELA) which consists of:

(a) an SMLA core;

(b) an SMLA ring;

(c) an outer ring composed of all local authorities which send more commuters to the core in question than any other core.

Using these definitions, Coventry consists of a core together with an inner ring and a discontinuous outer ring and is one of 128 MELAs in the United Kingdom (Figure 2.5). The SMLA/MELA approach provides, in theory, a more refined basis for urban study by distinguishing three types of urban area according to occupational and functional character. It remains, however, subject to all the criticisms of administrative definitions since it is based upon aggregations of local authority units. The indivisibility of these areas accounts for the irregular shape of many MELAs as the Coventry example shows.

Many countries now use functional criteria as the basis of official definitions of 'the extended city'. In the United States it is the standard metropolitan statistical area; in Canada the census metropolitan area; in Australia the census expanded urban district; in France the *agglomeration;* in West Germany the *Staadt* region; and in Sweden the labour market area (Simmons and Bourne, 1978). In each country, the criteria used are different, and the building blocks also vary but these initiatives represent positive steps by census authorities to define urban places on a more meaningful statistical and geographical basis.

An alternative approach to urban definition in the United Kingdom is possible with the release of the 1971 census data on a grid square basis. Grid squares are easy to map, and as all grids contain the same area of land, all counts are automatically density measures, and absolute numbers can be mapped directly. They are, moreover, unlikely to change over time and so will provide a solid basis for historical analysis (Rhind *et al.,* 1980). The availability of geocoded data enables urban geographers to move away completely from arbitrary administrative definitions of towns and cities and to build up pictures of the urban landscape using a wide variety of indices. For example, population density produces a clear differentiation among different categories of urban and rural areas in the English Midlands (Figure 2.6).

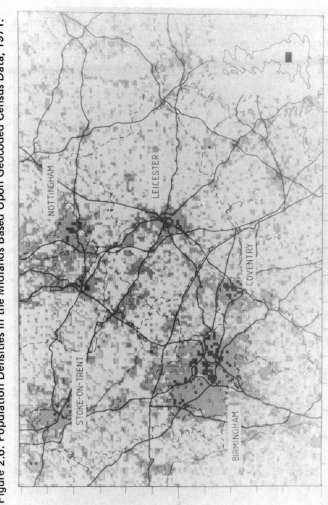

Figure 2.6: Population Densities in the Midlands Based Upon Geocoded Census Data, 1971.

Source: Rhind *et al* (1980), p. 10. Reproduced by permission of the *Cartographic Journal* and the authors.

The size and complexity of the modern city means that for most purposes, geographers have little choice but to rely upon secondary sources of data and so census authorities play an important role in shaping our views of urban geographical structure. For the most part, census enumerators tend to be conservative in their approach to defining urban areas. They cherish the boundaries of local municipalities because these municipalities are major data users as well as the source of much political power, and steer clear of the controversy and technical difficulties that attend more experimental aggregations. Recent technical advances in computing systems and in the technique of geocoding have made it possible for research workers to develop their own definitions and to create their own aggregations of urban units from census data. But until such techniques are in widespread use, and there is general agreement about the most appropriate criteria, the objective definition of towns and cities will remain a basic problem that must be faced by all urban geographers.

Conclusions

Cities exist in a number of ways. In one respect they are a set of images in the human imagination, in another they are precisely defined statistical and spatial units. This distinction introduces a major dilemma into urban geography. Analysis at the experiential level is difficult because images vary so widely, but it relates most closely to, and therefore has most meaning for, the residents of the city. Studies, however, of the objective city, while not free from definitional difficulty, are much more straightforward. They are likely to serve the needs of policy makers and planners for whom the city exists as an area on a map, but to be divorced from, and therefore largely irrelevant to, the urban realities as perceived by the population. No single solution exists at present to this dichotomy. Urban geographers are essentially 'outsiders' who view the city in a very different way to the 'insiders', the urban residents. Such fundamental differences of perception are in many ways central to the understanding and resolution of many contemporary urban problems.

3 THE GROWTH OF TOWNS AND CITIES

Rapid urban growth is a dominant feature of the development of most advanced economies. Its occurrence is so general, and its implications are so wide, that it is possible to view much of recent social and economic history in terms of the attempts to cope with its varying consequences. The rise of great cities and their growing areal influence initiated a change from largely rural to predominantly urban places and patterns of living that has affected most countries over the last two hundred years. Today, not only do large numbers of people live in or immediately adjacent to towns and cities, but whole segments of the population are completely dominated by urban values, expectations and life styles. From its origins as a locus of non-agricultural employment, the city has become the major social, cultural and intellectual stimulus in modern urban society.

Urban development is the process of emergence of a world dominated by cities and by urban values. It is important, however, to draw a clear and firm distinction between the two main processes of urban development; urban growth and urbanisation. Urban growth is a spatial and demographic process and refers to the increased importance of towns and cities as concentrations of population within a particular economy or society. It occurs when the population distribution changes from being largely hamlet and village based to being predominantly town and city dwelling. Urbanisation on the other hand is an aspatial and social process which refers to the changes of behaviour and social relationships which occur in society as a result of people living in towns and cities. Essentially, it refers to the complex changes of life style which follow from the impact of cities on society.

Historically, these two processes of urban development were interdependent. As people congregated in towns and cities, so profound and irreversible changes to their traditional ways of life took place. Moreover, because of their close association, the term 'urbanisation' was widely used to describe both the growth of towns and cities and the impact of cities on society. Today, the existence of two separate processes is recognised. One reason is that people can concentrate in space without experiencing any significant and immediate change in their pattern of living, as is the case in many of the rapidly growing cities in the Third World. The other is that with high speed transportation, telecommunications and the mass media, the inhabitants of even the most remote rural regions in advanced economies can participate in, and so will be influenced by, an urban culture. The terminology used here reflects a need to distinguish between the spatial and the sociological aspects of urban development. This chapter focuses upon urban growth; urbanisation is discussed in chapter 4.

Measuring Urban Growth

Urban growth takes place when the number of people living in towns and cities increases relative to the population as a whole. Within the limits of census definitions it is a simple concept to operationalise and measure. At any one time, the percentage of the population that is urban is given by:

$$\text{Per cent urban} = \frac{\text{Population living in towns and cities}}{\text{Total population}} \times \frac{100}{1}$$

and any increase in this figure constitutes urban growth. If it is assumed that the total population of most countries is rising or is at least constant, then urban growth will be determined by changes in the number of people living in towns and cities. Tisdale (1942) has pointed out that an increase in the urban population can occur in basically two ways; either the number of urban centres can increase, or the number can remain relatively constant and the population of centres can increase.

The proliferation of centres was the predominant process of urban growth throughout history, as the stock of urban places was continually added to in the form of new centres located in expanding settlement areas. This process of growth was widespread in the UK in the seventh century as towns were founded in cleared areas in Wales, Scotland and the North East, though it is illustrated most vividly in the nineteenth century in North America by the succession of settlements founded in the wake of the westward progression of the frontier (Borchert, 1967). Urban growth through increases in the size of settlements is a more recent phenomenon, and is associated with the enormous growth of metropolitan centres which has taken place in the last hundred years. The predominance of very large cities is indeed one of the most distinctive features of modern urban systems in both the developed and underdeveloped worlds. In 1900, 15 per cent of the world's population lived in cities of over 100,000 population; in 1980 this was 29 per cent.

Theories of Urban Growth

Some of the most fundamental questions in urban geography are concerned with the reasons why people choose to live in towns and cities, and the ways in which cities grow. Although there are important economies of scale and social benefits to be derived from living in close proximity to one's neighbours, drawbacks such as congestion, noise, pollution and lack of privacy suggest that the emergence of cities as the predominant settlement form is not necessarily a logical and inevitable development. Theories of urban growth seek to identify the forces which permit and encourage large numbers of people to concentrate in comparatively small areas in space. Two broadly contrasting viewpoints are prevalent in the literature, one underlining the importance of the economic prerequisites and imperatives for urban growth, the other emphasising the roles of the social bond.

Economic interpretations of urban growth lay stress upon the savings of assembly, production and distribution costs which may be achieved through concentration. They argue that the emergence and growth of cities are the consequence of the search for the most economical forms of settlement. In primitive economies based upon labour intensive agriculture, the population is typically arranged in small dispersed village communities, a pattern which gives maximum access to the land. Towns and cities will emerge where the level of agricultural production generates an annual food surplus. This critical development frees part of the workforce from agricultural employment and gives rise to a range of craft and trade occupations which cluster together in space so as to gain the maximum benefits of economies of scale and agglomeration. As city dwellers are by definition non-food producers, the level of agricultural surpluses remains a major constraint upon urban growth. Cities which achieve the greatest size are those which are situated in the most productive agricultural areas, or are located on major land or sea routes where they can draw upon the surplus product of a wide area. With the development of industry and an increase in agricultural productivity, access to external raw materials and markets for manufactured goods replaces the volume of food imports as the major determinant of urban growth. In an economy in which only a minority of the workforce is engaged in agriculture, the size and growth of cities is determined by the structure and organisation of industry.

Underlying this economic interpretation of urban growth is a set of relationships which are explained by economic base theory. At its most simple level, the urban economy can be reduced to two interdependent

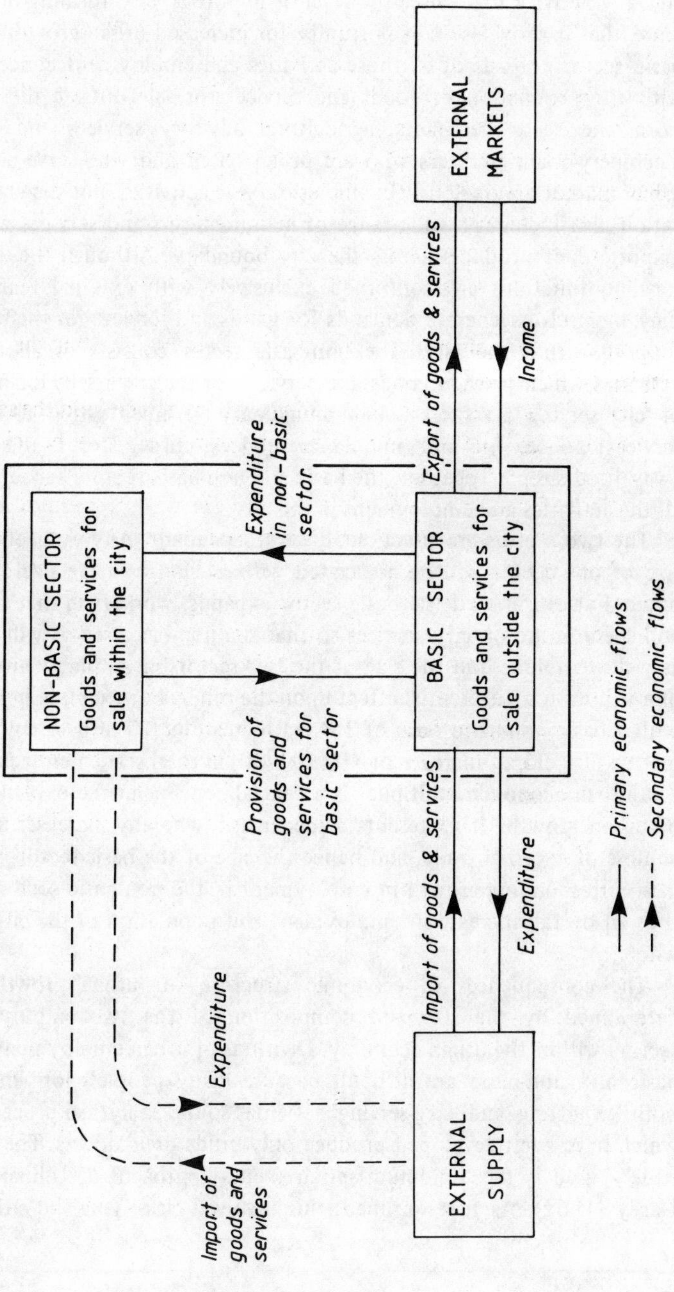

Figure 3.1: Basic and Non-basic Components Within the Urban Economy

sectors, the basic and the non-basic (Figure 3.1). Cities can only exist by buying in food and many raw material and product requirements, and any activity that raises the level of import is 'city forming' in the sense that it provides an opportunity for increased urban growth. The basic sector consists of all those activities and employment concerned with the production of goods and services for sale outside the city. Corn and seed merchants, agricultural advisory services and farm machinery manufacturers who are urban based and who serve a non-urban market are obviously examples of basic activities, but the classification also includes a wide range of manufacturers and services which 'export' their products across the city boundary. Although these city forming functions are concerned exclusively with external markets, they themselves generate demands for goods and services for their own support within the city. The non-basic sector consists of all those activities which provide goods and services for the city itself. Examples of 'city serving' activities include municipal government; street cleaning services; police, fire and ambulance services; corner stores; and take away food shops. Together, the basic and non-basic sectors account for all the activities and employment in the city.

The two sectors are functionally interdependent. Any change in the size of one sector will be associated with a change in the size of the other. For example, if the basic sector expands, workers in that sector will spend more on city services so that the non-basic sector will grow as well. Differences in the size of the two sectors mean that changes in one will have a differential effect upon the other. For example in a city with a basic:non-basic ratio of 1:3, an increase of 10 in basic employment will yield an increase of 40 (10 + 30) in total settlement activity. The urban economic multiplier is a central concept in the explanation of urban growth. It represents a mechanism whereby increases in the volume of external trade, and hence the size of the basic sector, result in a corresponding growth in employment in the non-basic sector, and thus an overall increase in employment and population in the city as a whole.

The consequences of economic structure for urban growth are determined by the size and composition of the two employment sectors within the urban economy. Divisions of urban employment into basic and non-basic are difficult because many people's jobs involve both 'exporting' and 'city-serving' activities so the analytical procedures which have been developed produce only crude breakdowns. The most widely used is the 'minimum requirements approach' of Ullman and Dacey (1962). As first outlined, this involved classifying the cities of

Figure 3.2: General Relationships Between Basic and Non-basic Components, and the Urban Economic Multiplier, With City Size.

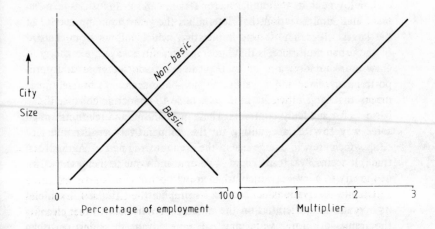

the USA into size groups and then examining the percentage of the total labour force which was employed in each of 14 categories of occupation. The lowest percentage recorded for any city in each group for the various occupations was assumed to be the minimum necessary to enable cities of that size to function. These minimum requirements were equated with the non-basic population and in a particular city the number of workers over this minimum figure was taken to represent the basic labour force. Empirical studies of US cities using this and related methods have identified the relative size of the basic and non-basic sectors (Figure 3.2). They show that the basic sector decreases in size as the urban population increases. For example, in a city of 10,000 people, approximately two-thirds of all employment is in basic activities, whereas in a city of 15 million, the figure is nearer one-quarter. A related finding is that the size of the multiplier similarily increases with urban size. It has a value of around 0.75 for a city of 200,000, but is nearer 2.00 for a city of six million.

Two important implications for the theory of urban growth follow from these findings. The first is that the larger the city, the less it is dependent upon the basic activities, and hence its links with surrounding markets, for its future growth. Whereas the volume of external trade is critical for small towns and determines whether they expand or decline, it is of secondary importance for the metropolis. Beyond a crucial population size, thought to be in the region of 250,000, growth is largely self-generated and is a product of the non-basic sector. Moreover, in view of the nature of non-basic activities, it is in fact mostly the level of employment in the public sector services and in industrial management and administration that determines the growth and prosperity of the largest cities. The second implication, which follows from the size of the urban multiplier, is that larger urban centres have the capacity to grow more rapidly: a small increase in the basic sector leads to proportionately large increases in the non-basic sector. Conversely, this means that large cities are more vunerable to economic collapse if the basic sector suddenly contracts. These mechanisms and relationships go some way towards accounting for the comparatively small number of cities which grew to a large size in the pre-industrial period. Agricultural trade, it seems, was insufficient in volume and value to increase the size of the city to a level at which urban growth became self-sustaining.

Despite reservations concerning its simplicity on the one hand, and its operational difficulties on the other, economic base theory provides the framework for a wide range of more advanced models of urban growth. The approach has been refined and elaborated by Pred in his analysis of city systems in advanced economies (1977). For Pred, the introduction of new or enlarged industry into a city triggers off two chains of reaction (Figure 3.3). The first is a set of initial multiplier effects which result in new construction, a growth in public transport and utilities, the expansion of service activity and employment and the enhanced likelihood of invention and innovation. The combined effect of new industrial employment and the initial multiplier effect will be an alteration of the city's occupational structure, and increase in population, and the probable attainment of one or more local or regional industrial thresholds. These higher thresholds, or larger markets, can support new manufacturing functions as well as additional plants or capacity in existing industrial categories. Once production facilities have been constructed in accordance with the new thresholds, a second round of growth is initiated, and eventually, still higher thresholds are achieved. This development is associated with secondary multiplier effects which lead to the introduction of additional industry and further invention and innovation which in turn opens up new external markets and so maintains and enhances the level of employment and population in the city. Such chains of reaction are circular, cumulative and self-reinforcing. They suggest that urban growth in advanced industrial economies is very closely related to the capacity of cities to innovate and change both their types of industry and the range and quality of the products which they manufacture.

Figure 3.3: The Circular and Cumulative Feedback Process of Urban Growth.

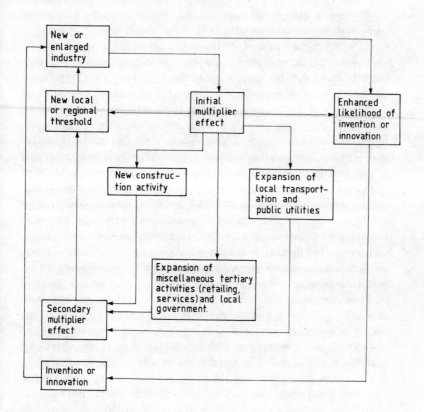

Source: Pred (1977), p. 90.

At first glance, the circular and cumulative feedback model appears to be far removed from the surplus agricultural product model of urban growth. The one is concerned with complex processes of urban expansion in advanced urban industrial society, the other seeks to account for the origin and growth of cities in primitive agricultural economies. In fact, these two approaches place common emphasis upon relationships and mechanisms which are explained by economic base theory. Each sees the city as a finite entity in geographical space which exists by exporting urban goods and services in exchange for those of external suppliers.

Regardless of whether the flow is in agricultural or industrial products, the critical factor in urban growth is, at least initially, the changing economic bond between the city and its outside suppliers and markets.

Although urban growth cannot occur without agricultural surpluses being available, an improvement in agricultural productivity may not of itself be the primary reason for the emergence of towns and cities. The development of urban centres in a wide range of economies and cultures suggests that urban life may be based upon fundamental interpersonal ties which encourage people to congregate together in space. Social explanations of urban formation stress the gregarious nature of human behaviour. They point to the complementary properties of individual human relationships such as male and female, mother and child, sender and receiver, giver and taker, and argue that such bonds introduce strong centripetal tendencies among human populations. Even small groupings offer security, defence and self-help, so they become increasingly attractive to non-group members, and with increased membership, the value of community benefits grows. Cities emerge when social institutions and mechanisms are developed which enable the population to live together in sizeable concentrations in space. Social organisation is, therefore, accorded primacy over economic developments as the independent variable in urban growth.

These arguments have been summarised succinctly by Adams (1960) in a study of urban growth in early Mesopotamia. He argued that 'the rise of cities was pre-eminently a social process, an expression more of changes in man's interaction with his fellows than in his interactions with his environment'. The novelty of the city consisted of 'a whole series of new institutions and the vastly greater size and complexity of the social unit rather than basic innovations in subsistence'. For Lampard (1965), society has evolved through four organisational stages each of which is associated with a distinctive settlement form. Particular emphasis is placed upon the 'primordial' since this represents the first achievement of a level of social organisation necessary to support and sustain village life. Improvements in agricultural productivity were necessary for urban growth, but the development of organised communities, capable of elementary agricultural management and a degree of social control, was crucial. This made possible trade outside the community, and rudimentary economic specialisation within, which in turn served to promote further settlement growth. The 'definitive' stage was reached when society achieved a level of organisation which was able to sustain formal religious, military, bureaucratic and political systems through which large concentrations of people could be controlled.

Cities, therefore, developed as organisational nodes; centres of accumulated social experience and knowledge to which they both add and export to surrounding settlements. The success of cities as definitive centres soon resulted, according to Lampard, in an era of 'classic' urban development which was characterised by a stable pattern of social organisation, and an interdependent network of hamlets, villages, towns and cities. This pattern was only broken in the last two hundred years when the technological and organisational capacities for unprecedented population concentration led to the growth of the large urban centres of the 'industrial' period. It will be seen from this summary that while the contribution to urban growth of economic productivity and technology is not denied, its importance is held to be contingent upon the achievement of a given stage in social organisation.

A more specific interpretation, which still stresses social relationships, was advanced by Meier (1962) in the form of a communication theory of urban growth. Meier analysed the nature of the bond between two individuals and suggested that the basic quantum involved is the transaction, the exchange or transfer of information between sender and receiver. Bond formation between individuals, he argued, is facilitated by geographical proximity and by the acquisition and retention of knowledge so that cities evolved primarily as a means of facilitating inter-personal communication. An important feature, indeed a major attraction of city life, is the time spent in public and professional as opposed to private and family life, so that shared symbols and experiences generate civic bonds which help to maintain and reinforce the cohesion of the city. Urban growth therefore takes place as cities develop a capacity to maintain and conserve information. The city is seen as an open communication system, resulting from, and held together by, a complex pattern of information exchanges.

Meier relates urban growth not to changes in economic product or to the size of the social group, but to developments in communications technology. Social exchange in early settlements was primarily achieved by face-to-face contact and, with people attempting to maximise their chances of social interaction by locating close to the central zone of conflux, cities were both densely populated and compact. They were dominated by a small elite of priests and politicians who, through public meetings and assemblies, controlled the distribution of community information. With the progressive introduction of handwritten records, printing, publishing and broadcasting, communications technology has complemented, and indeed now has largely replaced, face-to-face contact as the prime means of information dissemination. A different urban response and a subtly different form of social control can be identified at each stage of communications development. In general, densities have declined and peripheries have expanded, while power and control of the city has increasingly become associated with the ownership of newspapers, radio and television. Modern cities are defined not simply in physical terms but as social networks in space, created, maintained and manipulated by a wide range of communications media.

Theories of urban formation and growth have generated an extremely wide literature in urban geography. In many respects, however, the continuing speculation as to whether economic or social factors were historically paramount, has become something of a 'chicken and egg' debate. Urban growth requires both technical advances in agriculture and industry and the social capacity to implement these advances, and either by itself would appear to be insufficient. The rise of towns and cities is the outcome of a complex interweaving of social and economic relationships as even a superficial historical survey shows.

Stages of Urban Growth

In adopting a long-term perspective, it is apparent that the process of urban growth has undergone two major changes of pace (Figure 3.4). The first, known as the agricultural revolution, occurred in the Near and Middle East around the fifth millennium BC and is associated with the first emergence of identifiable towns and cities. The second, known as the industrial revolution, occurred first in Britain in the late eighteenth century, and led to the growth of the large modern metropolis. These revolutions separate what Sjoberg (1960) has termed the pre-agricultural, the traditional and the urban-industrial societies. They distinguish different technological environments each of which is associated with a specific settlement response.

The agricultural revolution is that name most commonly used to describe the stage in the development of civilisation when a system of organised agriculture replaced a nomadic and predatory way of life. An important consequence of the domestication of animals, and the cultivation of cereals, is that it led to significant increases in food production, over and above the levels necessary for basic subsistence, which could be used to support the establishment of settlements housing people engaged in non-agricultural activities. Urban centres thus emerged in the Nile, Tigris-Euphrates, and later in the Indus Valleys, and in the North China Plain, Southwestern Nigeria, Central America

Figure 3.4: The Growth of World Urban Population.

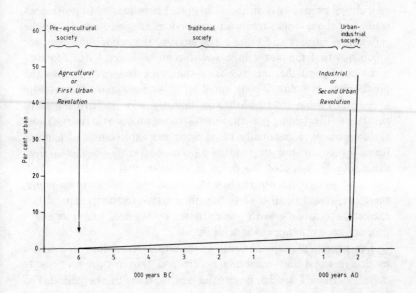

and the central Andes (King and Gollege, 1978). The major characteristics of these first centres of urban civilisation have been summarised
by Childe (1950). Although they would be considerably smaller than
many villages today, early cities were more extensive and densely
populated than any previous class of settlements. They were also
distinctive in morphological terms by virtue of their impressive array of
monumental public buildings. Early cities were closely tied to their
surrounding hinterlands for basic foodstuffs, but they also engaged in
'foreign' trade over surprisingly long distances so as to acquire essential
raw materials and religious artifacts. It was in social and economic
character and composition, however, that early cities represented the
most significant advance, for in addition to peasants who still worked
in agriculture, they included full-time craftsmen, transport workers,
merchants, officials and priests, all of whom were supported by surplus
agricultural product. Freed from the need to work in the fields, these
groups took part in literary, artistic and scientific activities, and naturally
assumed a privileged and elitist position in the social structure of the city.

Throughout the period of the traditional economy, a ceiling upon urban expansion was imposed by the availability of food surpluses which meant that the size of cities was closely determined by the level of agricultural productivity in the local area. Places located in prosperous regions or those whose transport connections enabled them to tap the agricultural surpluses of other areas, enjoyed the major advantages for urban growth. Examples include twelfth-century Venice, Milan, Florence and Geneva, and the port towns of the Baltic Hanseatic league. The productivity of industry constituted a further constraint as towns were in general unable to establish a sufficiently wide industrial base to generate self-sustaining growth. Manufacturing processes in the traditional economy were essentially based upon the exploitation of animate forms of power and the limited amounts of energy available from these sources restricted the range of materials that could be worked, and kept *per capita* output low. Cities in the traditional economy, therefore, found it difficult to buy their way into foreign agricultural markets, and urban growth was restricted by the need to devote a high proportion of resources to agriculture.

For successful exploitation, pre-industrial technology required a particular social and economic system and this in turn gave rise to major social and spatial similarities among cities of the period. Pre-industrial society was characterised by an all pervasive class system consisting of a small privileged elite, a vast lower class and a sizeable number of outcasts. Position in society was primarily determined by birth, and the family was the main unit of economic and social organisation. The elite exercised a monopoly of political power in the city and occupied key positions in religion, education and the bureaucracy. By industrial standards, the technology and economic organisation of the pre-industrial city were simple, and indeed commercial activities and manual labour were deprecated and shunned by the upper class. Manufacturing of goods was undertaken by craftsmen who performed all or most of the steps in the fashioning of an article. They were typically grouped into trades guilds which exercised wide control over recruitment, standards and pricing. Upward mobility within this rigid social hierarchy was minimal. The social order was perpetuated and maintained through an educational system which was primarily the preserve of elite males. Education involved traditional learning of the religious-philosophical variety which reinforced the ideals and values of the ruling class and helped them sustain their privileged position in society.

In identifying this social structure, Sjoberg (1960) has described the geographical arrangement of the pre-industrial city. Within the typical wall, the city tended to be divided into sections sealed off from one another by walls, moats and the like. The central area typically contained the prominent government and religious buildings and the main market. Clustered close to these major functions were the relatively luxurious dwellings of the elite, characteristically facing inwards and presenting an inhospitable blank wall to the streets. The distribution of population within the city from the centre was directly associated with power and wealth, the poorest living furthest out, often beyond the city walls. Within this geographical framework, the city tended to be subdivided on the basis of ethnic or occupational lines, but there was little in the way of contemporary territorial specialisation. Place of work was invariably identical to place of residence. The pre-industrial city, with limited means of circulating people and goods, was highly congested and lacked many of the amenities of contemporary urban life with respect to environment, sanitation and hygiene.

The second major change in the pace and nature of urban growth was associated with the complex set of economic and social transformations that are popularly known as the industrial revolution. Fundamental to urban growth were major improvements in agriculture, often linked with a change from subsistence to cash crop production, which provided the basic support for cities. Productivity was assisted by enclosure, farm rationalisation and technical innovation, and as *per capita* output increased, so the reduced demand for agricultural labour encouraged many workers to seek employment in urban occupations. Improvements in transport also assisted urban growth by enabling agricultural products and raw materials to be brought in from increasing distances.

Advances in agriculture were a necessary prerequisite, but rapid urban growth in the late eighteenth and early nineteenth centuries was a direct consequence of changes in the nature and scale of industry. The essential change was one of energy source with steam power from inanimate coal replacing the natural and animate energy forms of the pre-industrial period. For Geddes (1915), the late eighteenth century saw the emergence of a palaeotechnic technology based upon inventions like Darby's coke smelting process for iron (1709), Crompton's mule (1779), Cartwright's power loom (1785) and Watt's steam engine (1825). The palaeotechnic era was an age of coal and iron in which the economics of manufacturing and transport favoured the location of industrial production on, or very close to, energy sources. As a result, a class of coalfield cities emerged, first in Great Britain and subsequently in North

Western Europe and the north eastern USA, which produced the heavy manufactured products of the new industrial age.

The industrial revolution transformed Great Britain from a rural agricultural to an urban industrial economy. The pace of urban growth in the first half of the nineteenth century was unprecedented and unparalleled. Between 1801 and 1851 over nine million people were added to the population of England and Wales, but while the rural inhabitants (those living in places of less than 5,000 population) increased from 6.6 to 9.9 millions the town dwellers increased from 2.3 to 8.0 millions (Weber, 1899). Of the total increase, 64 per cent fell to the towns and cities. A population that was 26 per cent urban in 1801 was 45 per cent urban by 1851. By 1861, more people in England and Wales lived in towns and cities than lived in rural areas.

Much of the growth in the urban population at this time can be attributed to the increase in the number of large cities, as Hall *et al.* (1973) have shown. In 1801, London had about one million people and no other town in England and Wales had as many as 100,000. The biggest were Liverpool with 82,000, Manchester with 75,000, Birmingham with 71,000 and Bristol with 61,000. By 1851, the population of London was up to 2,491,000 (or 2,685,000 if the area of Greater London is considered); three other great provincial cities had over 200,000: Liverpool with 376,000, Manchester with 303,000 and Birmingham with 233,000. Four other cities had more than 100,000: Leeds with 172,000, Bristol with 137,000, Sheffield with 135,000 and Bradford with 104,000. Fifteen cities had over 10,000 population in 1801: by 1851 this had risen to 63.

The Palaeotechnic city was characterised by a highly centralised distribution of population and employment. Rapid population growth together with the absence of any means of mass transportation led to extremely high densities and extremely steep population gradients as the urban workforce clustered tightly around workshops and factories. In 1801, the most crowded districts of central London, totalling 1,045 hectares held some 425,000 people at an average density of 407 to the hectar (Hall *et al.*, 1973). By 1851 the most crowded area had extended somewhat to some 2,346 hectares and it housed 945,000 living at a density of 402 to the hectare. Manchester provided a similar illustration, as its five innermost statistical areas which covered 599 hectares in 1851 had an average density of 299 to the hectare.

Despite similarities in terms of location, size, density and rates of growth, the early industrial cities were very different in terms of socio-economic structure. Although grounded in the same palaeotechnic technology, regional specialisms in manufacturing — cotton in Lancashire, woollens in the West Riding of Yorkshire, metal goods in the West Midlands — and product specialisms among centres in these areas, together with the organisational requirements of individual techniques of production, led to basic differences in urban economies, social structure and municipal politics. For Briggs (1963) the first effect of industrialisation was to differentiate urban communities rather than to standardise them so that in any detailed analysis, the cities of the period must be treated as individual cases. The survey by Briggs of Victorian Manchester, Leeds, Birmingham, Middlesborough and London (as well as Melbourne, Australia) reveals the differing responses made by each city to the prevailing social, economic and political circumstances. The marked contrasts between mid-century Manchester and Birmingham account in part for the latter's lead in city development and improvement. Manchester's large-scale production in cotton mills and factories with huge machines tended by many operatives, contrasts with Birmingham's hundreds of small diversified workshop businesses each needing only a handful of skilled workmen. Further differences included Manchester's remote 'masters' and repressed and hostile 'hands', Birmingham's easier employer/worker relationships and wider scope for changing jobs: Manchester's municipal oligarchy, Birmingham's broader-based city administration and greater degree of democracy.

The industrial revolution initiated a progression to a modern world in which changes of industrial processes and products are the norm rather than the exception. By the last third of the nineteenth century, inventions such as the electric circuit (Seimens, 1850), the telephone (Bell, 1876), the power station (Edison, 1882), the oil well (Drake, 1856), the petrol motor (Daimler, 1883), the radio (Marconi, 1896) and many others were leading to the emergence of what Geddes (1915) termed neotechnic technology. This new technology and the industry it created was almost the exact opposite of that of the first industrial age as Hall (1966) has observed. 'Instead of heavy crude products, light and increasingly complex ones were developed; instead of coal, electricity; instead of the universal railway, increasing dependence upon the motor vehicle; instead of restriction, freedom of location through communications' (p. 24).

These changes in technology were accompanied by changes in the structure of firms and hence of the whole of industry. In the palaeotechnic era, industrial production in Britain was characterised by many small single function, single location factory firms. These were typically controlled by one man, a family or a partnership, and were financed out of profits, the savings of others, or modest borrowing from the local bank. With the demand for expanded production and more sophisticated technology, many of these factory enterprises were consolidated into national corporations typically engaged in many functions over many regions. Shannon (1931) saw this change occurring in Britain between 1844 and 1856 when joint-stock companies with limited liability received the general sanction of Parliament. To meet the needs of the new organisation of production, notably the need to co-ordinate widely scattered plants, a new administrative structure evolved. This involved both a horizontal division of management into specialised departments, and a vertical system of control to co-ordinate and connect departments. It gave rise to the head office, the responsibility of which was to organise, appraise and plan for the survival and growth of the corporation as a whole.

The tasks of administration and management were made easier by a series of technical inventions that occurred in the late nineteenth century. The list includes commercial shorthand (1837), electric telegraphy (1937), cheap universal postage (1840), the lift or elevator (1857), the typewriter (1867), the telephone (1876), the electric light (1880), the steel frame skyscraper (1875) and the development of adding and copying machines. Many of these inventions appear trivial in themselves but together, they created a controlled working environment and the means of storing, processing and duplicating information that was essential to the office function. From then on, the pattern of urban growth and structure was determined not only by the needs of industrial production, but by the requirements of industrial management and control as well.

For Hall *et al.* (1973), these technological and organisational developments initiated a period of urban growth in Britain that was characterised by contrasting trends in the geography of employment and population in the city (Table 3.1). In industry, the organisational split between production and management led to the spatial separation of plant-based manufacturing activities and office-based administration and control. Whereas the former expressed a preference for those locations which offered minimum assembly, production and distribution costs, and which were increasingly non-urban, control functions favoured city centre locations where they could ensure a high level of face-to-face interaction with a wide range of financial, insurance, advertising, legal and research organisations. This concentration of office jobs was reinforced by the expansion of the retail sector, and together, these increases in service occupations at the centre more than compensated for losses

Table 3.1: Stages in the Evolution of United Kingdom Cities, 1760-1980

Type of Economy	Dominant Modes of Transport	Forms of Corporate Structure	Trends Within the City	
			Population	Employment
Palaeotechnic manufacturing 1760-1880	Foot; Steam train	Factory	Centralisation	Centralisation
Neotechnic service 1880-1950	Tram, bicycle suburban railway; motor bus	Multi-department national corporation	Relative decentralisation	Centralisation
Managerial 1950-	Private car	Multi-divisional national and multinational corporation	Absolute decentralisation	Relative decentralisation

of manufacturing jobs to the periphery and beyond. The most pronounced feature of urban changes in population terms was out migration from the core to the edge of the city.

Although cities continued to grow in size, it was in areal extent that expansion was most dramatic. For example, between 1919 and 1936, Greater London increased in population from six to eight million, but it expanded in area five times. The outward spread of the city was facilitated and indeed encouraged by the development of public transport networks. The suburban railway, the electric tram and later, the motor bus, enabled people to live increasing distances from, and yet still commute to, their jobs in the centre. A dominant central business district and suburban sprawl reflect the two most powerful forces at work in the neotechnic city: centralisation of employment, and a relative decentralisation of population.

The post-war period in the United Kingdom has seen the emergence of a managerial economy characterised by a growing number of jobs in quaternary sector ideas and information activities. A major reason has been the appearance of divisional forms of corporate structure which feature several tiers of corporate control (Chandler and Redlich, 1961). In the multi-divisional national and multinational corporations, firms are decentralised into several divisions, each concerned with one product line, and organised with its own head office, while at a higher level, a group or corporate head office is created to co-ordinate the divisions and plan for the enterprise as a whole. A similar expansion of bureaucracy has taken place in the public sector, in local and central government and in health and welfare services, so that in 1971, 34 per cent of the jobs in England and Wales were in professional, managerial and administrative occupations. These structural developments have been accompanied by important demographic and social changes including a reduction in family size, increased longevity and an increase in the rate of household formation. The relative affluence of the post-war period is reflected in the high level of private car ownership and mobility, and by increasing demands for consumer goods, and new and improved housing.

Contrasting trends in the distribution of population and employment are apparent in most of the major United Kingdom cities since 1951. Over the last thirty years, redevelopment of the inner city, both for commercial purposes and for slum clearance, has caused a massive and sudden drop in central area population. Much of the overspill population passed directly into new housing developments in the suburbs, and the result has been a marked absolute decentralisation of population. Despite a continuing increase in the level of office activity in the central business district, the general economic decline of central areas contrasts with a rapid growth of service jobs in the suburbs, with the result that a relative, and in some cases an absolute decentralisation, of employment has occurred. These trends are almost the exact opposite of those responsible for the growth and form of the palaeotechnic city. UK cities are 'turning themselves inside out' with the jobs and population which were once in the centre, increasingly to be found in the suburban areas.

Urban Growth in the USA

The preceding review of stages of urban growth has been based upon the British case because it was the first country to experience an industrial and an urban revolution. Britain's lead in urban development was maintained throughout the nineteenth century so that by 1900, urban growth – measured in terms of the concentration of the population in urban places – had reached present-day levels (Figure 3.5). At that time, no other country remotely approached the level of urban growth in Britain, and only Belgium and Germany had more people living in towns and cities that in rural areas. In general, however, the later each country became industrialised, the faster was its urban growth. The change from a population with ten per cent of its members living in cities of over 100,000 to one in which 30 per cent lived in such cities

Figure 3.5: Urban Growth Curves for Selected Countries, 1851-1981.

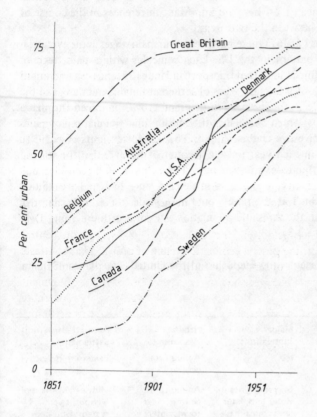

took about 79 years in England and Wales, 66 in the USA, 48 in Germany, 36 in Japan and 26 in Australia (Davis, 1965).

Urban growth in most of the countries of Western Europe was a corollary of industrialisation, and the expansion and evolving character of towns and cities was primarily a response to changes in the scale and nature of industry. Although the cities of the north-eastern seaboard went through similar stages of development, the pace of urban change in the USA was determined as much by the opening up of lands in the west for settlement as by industrial changes in the east. Between 1790 and 1920, the frontier of permanent settlement progressed across 2,000 miles of territory between the Appalachians and the Pacific coast. This continental dimension is a feature of the urban geography of the USA which is totally absent in a Europe divided into a large number of small countries. Fundamental contrasts of geographical scale and historical background mean that there are important differences in the course of urban growth between the two areas.

For Borchert (1967) there have been four main stages in the evolution of the urban pattern in the US. Each coincides with a major techno-logical era, defined in terms of transport and industry, and is characterised by the growth of particular types of settlement in different areas of the country (Table 3.2). The pattern of urban growth is traced through a series of maps which show the distribution and population of cities according to five size classes. (Figure 3.6). The lower limits of each size class increase in each year to take account of the fact that the scale of urban growth increases through time. For example, to be considered as a fourth order city in 1790, a settlement only had to have 15,000 inhabitants which at the time would have made it a sizeable place. In 1960, however, the threshold population size for fourth order cities was 250,000 (Table 3.3).

Table 3.2: Stages in the Evolution of the United States Urban System, 1790-1960

Epoch	Major Innovations	Urban Response	Areas of New Settlement
Sail and wagon, 1790-1830	—	Agricultural centres	Eastern midwest
Steamboat and iron horse, 1830-70	Steam engines in water and, later, land transport; iron technology	Agricultural centres; river towns, later, railroad towns; small scale manu-facturing centres	Midwest; Mississippi-Missouri river system; Gulf coast
Steel and rail, 1870-1920	Steel technology in industry and land transport; electric power	Agricultural centres; mining centres; large scale manufacturing centres	High plains; West coast; South
Auto-air-amenity, 1920-60	Internal com-bustion engine; tractor, car and aeroplane	Service centres; recreation and resort centres	Gulf coast; South; Southwest

Source: after Borchert (1967).

Figure 3.6: The Changing Pattern of Urban Settlement in the United States, 1790-1960.

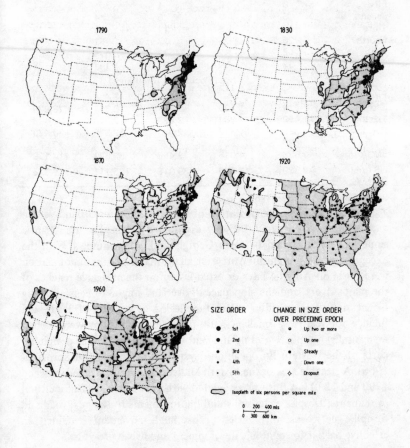

Source: Borchert (1967), pp. 301-23. Adapted (reprinted) by permission of the American Geographical Society.

Table 3.3: United States: Number of Cities and Total Population by Size Order, 1790-1960.

Size Order	1790	1830	1870	1920	1960
			Number of Centres		
First	0	0	1	1	1
Second	3	3	6	4	6
Third	8	8	14	16	19
Fourth	20	29	33	51	70
Fifth	8	12	37	75	82
Total	39	52	91	147	178
			Total Population (thousands)		
First	—	—	2,171	8,490	14,760
Second	514	1,120	3,301	10,364	28,826
Third	499	784	3,627	13,918	26,493
Fourth	530	1,812	2,533	12,829	30,473
Fifth	95	300	1,826	6,972	12,647
SMSA total	1,638	4,016	13,458	52,573	113,199
US total	3,929	12,866	39,818	105,711	179,323

Source: Borchert (1967), Table 2, p. 315.

Urban settlements in 1790 were essentially confined to the Atlantic slope and coastal plain between Georgia and Massachusetts. With the exception of Pittsburgh and Worcester, all the major centres of population were ports on Atlantic bays or estuaries, or on the navigable reaches of the major rivers, and the slow pace of overland wagon or coastal sailing meant that the individual cities functioned as centres of trade and finance for relatively small agricultural hinterlands. They were in fact mercantalist outposts of England and could be regarded as peripheral parts of the Western European city system at that time (Luckermann, 1966). A sizeable area in the eastern Midwest was homesteaded between 1790 and 1830 but this was associated with only small scale agricultural settlements. Despite extensions and improvements to the post road and turnpike networks, difficulties of overland movement remained an effective barrier to economic development and urban growth.

Changes took place in this urban pattern in the next epoch as a result of settlement in states bordering on the east bank of the Mississippi, and the development of the steamboat and the 'ironhorse'. Although steamboats had been successfully used as early as 1810, the real build-up of steamboat tonnage on the Ohio-Mississippi-Missouri system began in the 1830s, and the main period of increase in the tonnage of general cargo vessels on the Great Lakes also began in the 1830s and 1840s.

Rail mileage likewise grew rapidly after initial development in 1829 so that there were 9,000 miles of track by 1850, and 53,000 by 1870. At first, improvements in water transportation led to the emergence of river and lake towns, particularly New Orleans, St Louis, Cincinnati, Detroit and Cleveland, and the growth of many Midwest centres, especially Chicago, was subsequently stimulated by the expansion of the rail network. Boom centres also appeared in the coalfield areas of the Appalachians in response to the increasing demands for coal and iron which were generated by the early palaeotechnic economy. On the Atlantic coast, New York emerged as a first order centre, but Charleston, South Carolina and many places in New England, which were comparatively remote from the emerging industrial heartland, began a prolonged relative decline.

The course of industrialisation was reinforced between 1870 and 1920 by the appearance of abundant quantities of low priced steel, and by the rapid development of manufacturing. With the refinement of steel technology, steel rails replaced iron on both newly built and existing lines, and heavier equipment, more powerful locomotives and a standardisation of gauges, permitted increased speeds and the coast-to-coast shipment of goods. The major consequence of a drastic reduction in the frictional effects of distance was a widening of market areas, which provided for the large-scale growth of manufacturing centres in the East. During this period, nearly all the great metropolitan commercial centres of the Midwest and Northeast, while establishing themselves as major industrial cities, retained their positions or advanced one level in the urban hierarchy. Conversely, many inland river towns and a number of small manufacturing centres in New England declined in size order.

In the West, the settlement of the Great Plains was associated with the growth of large cities but these were widely scattered owing to the extensive nature of agriculture in the region. Urban growth in the Pacific Cordilleras developed in conjunction with the exploitation of accessible and usable minerals, or of desert oases, but only Salt Lake City and Denver had grown to metropolitan status by the end of the epoch. The 1870-1920 period was one of significant urban growth on the Pacific Coast. By 1920, Seattle, Portland and Los Angeles had emerged as sizeable urban concentrations while San Francisco maintained second order status. In the Central Valley of California, the growth of specialised agriculture resulted in the emergence of a number of fifth order processing and service centres.

The fourth technological era dates from the introduction of the internal combustion engine in transportation and related technology.

For Borchert, the internal combustion engine not only increased individual mobility and so further extended urban fields, but by putting the farmer on the tractor, it multiplied the land area he could work alone, initiated a revolution in farm family size and accelerated the urbanisation of 'rural' America. These changes coincided with the emergence of a neotechnic economy in which over half the workforce was employed in service activities, and in which individuals had increasing amounts of leisure time at their disposal. In place of changes in technology and industrial energy, urban growth in the auto-air-amenity era reflected the demands of mobile, affluent, white collar and leisure-rich population.

By the start of the epoch, the present pattern of settled areas had been established and subsequent changes took place within that pattern. The most conspicuous was the growth of employment in places where there were already large concentrations of people to be served, and with growth leading to further growth, the population progressively arranged itself into small numbers of metropolitan centres. In 1960, 31 per cent of the US population lived in SMSAs of over one million population. Within the major urban agglomerations, central areas in general grew less rapidly than the suburbs so that relative centralisation of population and employment became an increasing feature of urban change at the metropolitan level. Boom towns in the 1920 to 1960 period included automobile manufacturing centres in the Midwest, oil-field cities in Kansas, Oklahoma, Texas and the Gulf coast, while the influence of amenities upon the location of industries and people was reflected in the growth of new metropolitan centres in Florida, the Southwest and in Southern California. All the cities on the West coast, except Seattle, moved up one or two ranks in size order. The cumulative effect of urban growth processes in operation since 1790 is a highly centralised pattern of settlement in which 95 per cent of the US population live either in central cities or within daily commuting distance (Figure 3.7). Areas beyond these daily urban systems are either unpopulated or experiencing population decline.

The spectacular growth of the largest SMSAs in the auto-air-amenity epoch was responsible by 1960 for the emergence of a super-metropolitan region of 37 million people between Boston and Washington (Figure 3.8). The unprecedented scale and distinctive character of this urban development was recognised by Gottmann (1961) who, in naming it 'Megalopolis', argued that it was in the north-eastern seaboard of the United States that the ancient Greek dream of a super-city was finally realised. In *Megalopolis: the Urbanised Northeastern Seaboard of the*

Figure 3.7: The Contemporary Urban Pattern in the United States.

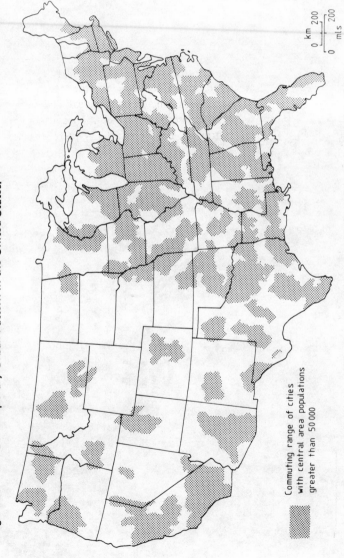

Commuting range of cities
with central area populations
greater than 50 000

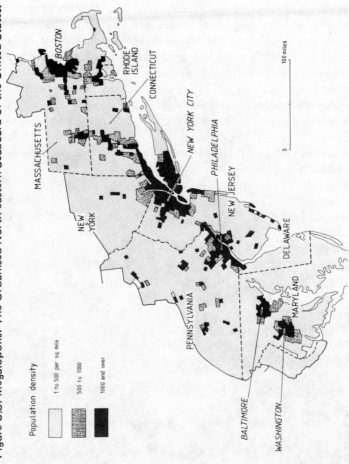

Figure 3.8: Megalopolis: The Urbanised North-eastern Seaboard of the United States.

United States, Gottmann painstakingly analysed the growth and attributes of the region. Alone among American urban areas, Megalopolis benefited from growth processes operating in all four preceding technological eras. It arose as a grouping of the main seaports, commercial centres and industrial activities in the United States, and to a large extent this maritime façade feature is still carried on. Similarly, the manufacturing function has never stopped and, indeed, a considerable expansion has taken place in the post-war period, of heavy iron and steel, chemical and metallurgical industries. Growth in the last half-century has been a consequence of the attraction to Megalopolis of a disproportionate share of the nation's financial, political and industrial decision making functions. As a consequence, it represents the wealthiest and best paid concentration of people in the world, and enjoys a social and cultural pre-eminence reflected in an exceptional concentration of great universities, libraries, publishing houses and centres for the visual and performing arts. What makes Megalopolis distinctive is not, however, simply its population size and areal extent though these are remarkable enough, but the degree to which the individual SMSAs interact and interdepend. The commuting areas are not merely contiguous, they overlap and interlock in complex ways so that many areas are influenced by more than one city. Secondly, the area is held together by a functional network consisting of a highly varied exchange of people, goods and ideas. For Gottmann, Megalopolis is rather more than a coalescence of adjacent urban centres, it is a supra-metropolitan system representing the cradle of a new order in the organisation of inhabited space.

Since 1960, 'megalopolis' has lost its distinctive North American connotation and has become a general term for the largest polynuclear urban systems. If the minimum population is set at 25 million, there were six megalopolises in the world in 1976: the north-eastern seaboard of the US ('Bos-Wash'), Chicago-Pittsburgh ('Chippits'), Osaka-Kobe, Japan, the 'axial belt' of England, the megalopolis of northwestern Europe and the Shanghai Urban Constellation, China (Gottmann, 1976). To these cases, Gottmann further suggests that three others may soon be added. These are the Rio de Janerio-Sao Paulo complex in Brazil, the Milan-Turin-Genoa triangle in Mediterranean Europe and the San Francisco-San Diego corridor in California ('San-San'). Megalopolises, whether existing or emerging, are in the most massive and impressive contemporary urban form. They represent the most recent stage in a major process of population concentration and urban growth that began with the emergence of the industrial city, almost exactly two centuries ago.

Urban Growth in the Third World

Despite differences in time and spatial scale, urban growth in the UK and the USA, as well as in most of Western Europe, Canada and Australia, was largely a product of industrialisation. The introduction of factory systems of production required a concentration of labour which in turn generated demand for retail and service activities, themselves increasing population concentration and the demand for industrial products. Secondary forces and processes of population growth, especially rural-urban differences in birth and death rates, and migration, both internal and international, increased the numbers living in cities in relation to the population as a whole so that growth, once initiated, was cumulative and self-reinforcing. The geographical pattern of urban development in terms of central business districts, suburban sprawl and declining inner areas, reflects both the changing locational preferences of industry, and the differential prosperity of the labour force.

In contrast to Western experience, the rise of cities in the Third World must be viewed as a product of diffusion rather than of indigenous urban growth. Sizeable cities existed outside Western Europe and North America at the turn of the century, but these were largely ports or mining centres linked to Western colonialism. They formed in each country the upper layer of a dual economy which was superimposed upon, but only weakly linked to, the traditional village-based agricultural system. The result was that the urban populations of Africa, South and East Asia were comparatively small in percentage terms (Table 3.4). Today, these figures are considerably higher, and as a consequence of the large populations of China, India, Egypt, Nigeria and Brazil, half of the world's urban population now lives in Third World cities. Urban growth in the Third World is taking place in countries with lower levels of economic development, at a far more rapid pace, and for different reasons, than was the case in the West. Natural increase and massive immigration are primary factors in this process: industrial development and its associated employment opportunities is a secondary cause.

The most important contrast between traditional Western and contemporary Third World urban growth processes is in terms of fertility and mortality. In the nineteenth century, UK city death rates were extremely high as a consequence of poor sanitation, poor housing and poor nutrition and as birth rates were significantly lower than in rural areas, the city was an area of natural population decline. Urban growth was primarily a product of rural-urban migration. Conditions in

Table 3.4: World and Regional Urban Population Statistics 1925-75

	Urban Population (millions)			Urban Population (percentage)			Million — Cities			
							Number		Percentage of population resident in	
	1925	1950	1975	1925	1950	1975	1960	1975	1960	1975
World total	405	702	1,547	21	28	39	109	191	9	13
North America	68	106	181	54	64	77	18	30	26	33
Europe	162	215	318	48	55	67	31	37	17	19
USSR	30	71	154	18	39	61	5	12	6	10
East Asia	58	99	299	10	15	30	23	45	8	13
Latin America	25	67	196	25	41	60	11	21	15	22
Africa	12	28	96	8	13	24	3	10	2	6
South Asia	45	108	288	9	15	23	16	34	4	7
Oceania	5	8	15	54	65	71	2	2	25	27

Source: UN (1974), pp. 36, 63, 64.

the cities of the non-industrialised countries are not, however, as hostile to reproduction, and fertility levels are little different to, and may even exceed, those in surrounding rural areas. One reason is the general level of health because the major cities are the places to which medical and scientific techniques, expert personnel and funds from advanced nations are first imported and where most people can reach them at least cost. The other is that economic improvements, public welfare, international aid, subsidised housing and free education make the penalties for having children less than they once were. They are reinforced by policies designed to give priority to housing large families, to maintain maternal and child clinics, and to discourage labour force participation by married women. Many of these factors also account for the major advances in death control which have been achieved in many Third World countries in recent years. Mortality levels are still high in Western terms but are lower than in surrounding rural areas. High birth rate and low death rate means that Third World cities are characterised by large natural population increases and this provides a major, but still only a partial, explanation of rapid urban growth.

Natural increase is reinforced by migration. Indeed this was the primary component of growth in many Third World cities during the late 1950s and 1960s. Migration is in part a response to the opportunities provided by the city in terms of jobs and standards of living but the high levels of urban unemployment suggest that this cannot be the only consideration. For Berry (1973), an additional factor is the role of Third

World cities as centres of social and political change, a feature which amounts to 'new centrality'. The Third World city is thus a symbol, drawing in massive immigrant streams, especially of young men, from overcrowded rural areas, only to find rural poverty replaced by urban poverty.

Despite differences between Western and Third World processes, the consequence of urban growth — an economy dominated by a small number of large cities housing the majority of the population — appears to be the same. Contemporary levels of urban development in the Third World still lag considerably behind those in the West although individual cities such as Shanghai, Cairo, Mexico, Calcutta, Buenos Aires and Sao Paulo, already exceed five million population. The number of million cities in the world almost doubled over the period 1960 to 1975 and almost all of these new million cities were in Third World countries. In fact by 1975, more than half of all million cities were in less developed regions of the world. The growth of megalopolis and the evidence of continued and accelerating centralisation processes in many Third World countries raises important questions concerning the most likely distribution of population in the long term. Some possible indications are provided by observations on the contemporary American city which suggests that the urban pattern is being superseded by new and essentially different forms of human settlement.

The Urban Future

It is being increasingly debated among urban geographers whether the processes of population concentration which gave rise to the urban present, will continue in the future. One reason for this concern is the shift in employment from placebound primary (extractive) and secondary (manufacturing) into geographically flexible tertiary (service) and quaternary (information and ideas) sectors; another is the growing importance of telecommunication services as media of social and business exchange; while a third is the rapid ageing and decay of many urban environments. Attempts to incorporate these considerations in forecasts of the form and function of future urban systems are necessarily speculative since the science of prediction is at best imprecise. For this reason, research workers use synoptic models, as a means of ordering their speculations. Such frameworks are known as scenarios, hypothetical sequences of events which highlight causal processes and relationships. Scenarios possess a number of methodological advantages but do not lead to single views of what is to come, rather, they generate complex sets of alternative futures based upon contrasting processes of development.

In view of this close association with 'alternatives' it is not surprising to find that spatial scenarios are of two general kinds. One stresses centripetal processes that will lead to further population concentration and urban growth, the other supposes population decentralisation and dispersal. The former generates an image which is different in scale, but not in kind, to the urban present. For example, one prognosis is that 'the USA in the year 2000 will probably see three gargantuan mega-lopolises, Boston-Washington, Chicago-Pittsburgh, and San Francisco-San Diego, which will contain over half the US population' (Kahn and Weiner, 1967, p. 61). Another is that twelve great concentrations of megalopolises, interconnected by elongated strips of settlement, will characterise the US by the year 2060 (Doxiadis, 1966). Alternatively, a radical transformation of the contemporary urban situation is possible if centrifugal urban processes dominate. This scenario is based upon the supposition that the locational constraints imposed upon industry and population in the post-industrial economy are fundamentally different to those of preceding technological eras. Changing job opportunities combined with 'frictionless' space and the attractiveness of amenity environments could lead to a wholesale dissipation of existing urban clusters by the end of the century (Berry, 1973).

Fundamental to the debate are the roles and impacts of transport and telecommunications technologies. Evolutionary models underline the continuing primacy of physical movements and the continuing expansion of urban commuter fields, enabling cities to extend their influence over a wider area. This progressive reduction in the frictional effects of distance and the corresponding increase in urban centrality has been the dominant mechanism behind urban growth in the past as Borchert's (1967) 'stages of urban growth model' has emphasised (see Table 3.3). Each succeeding transport technology has extended urban influences further so that cities dominate dependent regions of increasing size, and in areas like the north-eastern seaboard they have coalesced and combined as elements within a multinodal megalopolitan system. Although future transport systems will further reduce the effect of distance, the introduction of teleconferencing, data and facsimile transmission services is rapidly removing the need for daily access to central cities, indeed to any particular location. Business conferences and social meetings can now take place remotely without the need for participants to congregate in the same physical place so that much of the rationale for the city, as a point of maximum accessibility, is removed

(Clark, 1979). One of the most important forces contributing to contemporary urban growth is, therefore, the erosion of centrality by time-space convergence. In essence, this scenario amounts to an addition to Borchert's model in the form of an entry for the telecommunication-computer epoch. New telecommunications services eliminate time and overcome space in such a way as to make possible an urban civilisation without cities.

The most important geographical implication is the possibility that the urban population could relocate on a massive scale. A comparatively small percentage of the workforce, notably those employed in extractive activities and in heavy, raw material oriented manufacturing, are tied by their jobs to a particular location, and the majority already enjoy a great deal of geographical flexibility. Moreover, increasing personal affluence, and a reduction in the duration of working life and the working week, mean that recreational and amenity environments, which offer escape from pollution and congestion, could become the most attrative. For Berry (1973), there is already evidence of this trend in the US. Settlement patterns are spreading broadly over North America, and densities are increasing most rapidly in localities where climate and landscape are the most pleasant. In part, this represents a continuing reaction to the increasing racial and economic problems of central areas which fed the suburbanisation processes of the early post-war decades, but the scale and significance of this dynamic is altogether more radical and far-reaching. What is involved is continental dispersal rather than local decentralisation. Paradoxically, the remote hills, lakes, forests, deserts and shorelines of South Texas, southern California, southern Arizona and southern New Mexico which are now in demand, were environments that were historically ignored. Rather than a consolidation of the existing urban pattern, this points to a revolutionary and rapidly advancing inversion of the urban geography of the US by the end of the present century.

Conclusion

This chapter has focused upon theories and patterns of urban growth. Both economic and social interpretations of settlement formation and expansion have been advanced, the former stressing the savings of assembly, fabrication and production costs than can be achieved through population concentration, the latter emphasising security and community benefits, and the opportunities provided for inter-personal communication. Despite ancient origins, the city remained an exceptional settlement form until the late eighteenth and early nineteenth century when an urban-industrial take-off occurred in Great Britain and parts of North-western Europe and North America. The subsequent course of urban growth reflects both changes in the structure of industrial capitalism and in transport technology so that a number of distinct urban stages are discernible. These are different in character and time-scale in Great Britain and the United States because of history, the characteristics of diffusion of industrial technology, and of the continental dimension to North American urban development. Urban growth in the Third World is in part an adoption of Western urbanism though it is characterised by a distinctive and powerful demographic dynamic. Today, over 40 million people live in the Bos-Wash megalopolis of the North-eastern Seaboard of the United States. Conversely, the agricultural peasant village remains the dominant settlement form throughout most of the Third World.

Urban growth at the world scale is increasing rapidly. If Western experience is any guide, then dramatic increases in the percentage of the population living in Third World towns and cities may be expected over the next fifty years so that the great cities of the early twenty-first century will almost certainly be found in Africa, Central and Latin America and the Far East. It would be wrong, however, to assume that megalopolis, though it is the product of many decades of population concentration, is necessarily the ultimate settlement form. Decentralisation followed by dispersal processes associated with the social and spatial changes implicit in the transition from an industrial to a post-industrial economy suggest that new and essentially different forms of settlement are emerging in the West. Urban geographers are conscious that their viewpoints and perspectives are both time- and place-bound. The contemporary settlement pattern in the West is the most recent stage in a long, continuing and ever changing process of urban growth: there is no assurance, however, that it will be necessarily replicated elsewhere.

4 THE URBANISATION OF SOCIETY

Having examined the underlying causes of urban formation and growth, it is now appropriate to explore the effects upon people of living together in towns and cities. As the population concentrates in relatively small areas in space, so the forms of social and economic structure and organisation which are appropriate for rural living break down, and are replaced by new patterns and relationships more suited to urban needs. At first, these changes are restricted to and are experienced by those actually resident in the city, but over time, they diffuse to and are adopted by those living in rural areas, so that the whole of society becomes dominated by urban values, expectations and life styles. This process of behavioural and relational change is known as urbanisation.

That important socio-economic differences characterised rural and urban areas was recognised by many nineteenth century observers. Writing in 1899, Weber traced the growth of cities over the preceding one hundred years, and explored, using census data, the major socio-economic correlates of urban life. He found that cities contained a larger proportion of women and of foreign-born than did the rest of the country, and had divorce rates three to four times those in rural areas. There was a regular increase in the proportion of women to men as one ascended from small to large cities and a parallel decrease in the married in each group. The excess of women was among the city-born rather than the immigrants. Fewer females than males were born in the city compared with the country, but infant mortality was found to be highest for males. Violent deaths affected men principally because city occupations were more dangerous to the health of males than country occupations, as were crime, vice and excesses of other kinds which shorten life.

Weber also examined the physical and moral health of the city through an analysis of birth and death rates, and found that the towns-man was on the average shorter lived than the countryman. He was also less healthy and less vigorous and capable, both physically and mentally. City life produced fewer of the most severe physical infirmities than did the country, but it did favour the increase of insanity. Suicide rates were also higher in cities, a characteristic Weber saw as one of the penalties paid for progress, resulting in failure in the struggle for exist-ence. The statistics for crime showed cities to have several times the rates of the country, and also to be characterised by vice and illegitimacy. To these crude statistical differences, others added vivid descriptions of the social conditions in the city, none more powerful than Mayhew's 4-volume *London Labour and the London Poor* (1851-62) and Booth's massive *Life and Labour of the People of London,* published in 17 volumes between 1889 and 1903. Together, these works emphasised in both quantitative and qualitative terms the enormity of the social and economic gulf which at the time separated town and country.

Urban Ways of Life

Present-day theories about the social and behavioural consequences of urban growth trace their origins to the writings of sociologists who observed rural and urban life in the nineteenth century. In developing an understanding of societal change, social analysts assumed that the industrial metropolis marked a fundamental cultural divide. The city represented a novel scale and quality of socio-economic organisation that was an inherent and inevitiable departure from that long established in rural areas. Early explanations of urban life were thus couched in terms of 'so called theories of contrast' (Reissman, 1964, p. 123). Despite the wide differences in terminology, these recognised two different types of society, a traditional rural and a modern urban, each characterised by a different social and behavioural order (see Table 4.1).

Studies of urban-rural contrasts are well represented by the work of Tönnies (1887). From observations of contemporary German cities, Tönnies claimed to see a new pattern of social organisation evolving. He argued that rural life took place within the framework of *gemeinschaft* (community), whereas urban life was characterised by *gesellschaft* (society). In the former, the basic unit of organisation was the extended family or kin-group, within which, roles and responsibilities were defined by traditional authority, and social relations were instinctive and habitual. Co-operation was guided by custom. In the latter, these close, instinctive and established patterns were replaced by the formalised, contractual, impersonal and specialised relationships of the *gesellschaft*. City life was characterised by competitive bidding for labour as one of several factors of production in the market place and by the increasing influence, in human relationships, of one's professional peers. With the decline in the importance of the family, social interactions were re-organised on the basis of rationality and efficiency rather than tradition.

This basic conceptualisation of rural-urban differences was elaborated

Table 4.1: Polar Distinctions Between Pre-industrial and Urban-industrial Society.

	Pre-industrial Society	Urban-industrial Society
Demographic	High mortality, fertility	Low mortality, fertility
Behavioural	Particularistic, prescribed; individual has multiplex roles	Universalistic, instrumental; individual has specialised roles
Societal	Kin-group solidarity, extended family, ethnic cohesion; cleavages between ethnic groups	Atomisation; affiliations secondary; professional influence groups
Economic	Non-monetary or simple monetary base; local exchange; little infrastructure; craft industries; low specialisation	Pecuniary base; national exchange; extensive inter-dependence; factory production; capital intensive
Political	Non-secular authority; pre-scriptive legitimacy; inter-personal communications; traditional bases	Secular polity; elected government; mass media participation; rational bureaucracy
Spatial (geographical)	Parochial relationships; close ties to immediate environment; duplication of socio-spatial groups in a cellular net	Regional and national interdependence; specialised roles based upon major resources and relative location within urban-spatial system

Source: Berry (1973), p. 13.

by Durkheim (1893). Durkheim saw the increasing division of labour as an irreversible historical-biological process involving the development of human civilisation from a segmental to an organised form. The segmental society was based on blood relations comprising a succession of kin-groups but with modernisation, these small units were grouped into larger aggregates. One result was the formation of territorial states, another was the occupational organisation of society, with individuals being classified according to the nature of the social activities they performed. Additional facets of pre-industrial and urban industrial society were illustrated by Simnel (1903) and Weber (1920) so that together, these social analysts laid the foundations for a general explanation of rural-urban differences. For Berry (1973), the 'conventional wisdom' they produced 'centred on the idea that there was a replacement of all encompassing primary social relationships touching on all segments of a common life experience, based upon sentiment, custom, intimate knowledge and hereditary rights, by impersonal secondary relationships based upon specialisation' (p. 12).

In the last fifty years, this view of rural-urban differences has been supplemented and elaborated by numerous workers, of which the best known and most influential were those belonging to the Chicago School of Human Ecology. Park (1916), the founder of the school, had proposed that the best method of studying the new urban ways of life was for the social scientist to go out and undertake exploratory studies in his or her own city. Unparalleled opportunities for field observation were provided by Chicago which, in the first two decades of the twentieth century, was an early industrial metropolis, expanding rapidly through immigration and exhibiting all of the stresses and tensions associated with explosive growth. On this basis, a large volume of research developed which provided a wealth of empirical data with a firm grounding in ecological theory. One of the last members of the school, Louis Wirth, was responsible for drawing all this work together in a seminal essay entitled 'Urbanism as a Way of Life' (1938).

Wirth set out to discover the forms of social action and organisation which typically emerge when people congregate together in cities. Specifically, he identified the three dominant characteristics of the city as being its large size, its dense population concentration and its heterogeneous social mix. Wirth began with size as the principal ecological characteristic of the city from which he next deduced propositions about urban society, in turn using these as a basis to deduce other propositions about the personality of urbanites. In this way, he synthesised the elements proposed by previous writers into a coherent theory which contained social structural, cognitive and behavioural components.

For Wirth, the size of the social group determined the nature of human relationships. Increase the number of inhabitants in a community beyond a certain level and the possibility of each member of the community knowing all the others personally is reduced. Moreover, urban dwellers do not become involved with one another as total personalities but in specialised segments, interaction being for definite and instrumental reasons. Under these circumstances individuals will tend to form only weak links with others so that the close bonds of family and neighbourliness, present in folk cultures, gives way to differentiation, specialisation and symbolism. Formal methods of social control replace informal methods, and indirect modes of communication replace personal contacts. The value of social relationships will be measured in monetary terms, and will be manipulated as a means of achieving one's own ends. The effect will be that the individual will come to count for little and will be subsumed as an anonymous member of a social group, responding to institutionalised codes of behaviour.

Lacking the security provided by familiar norms and sanctions, the individual will feel a sense of personal disorganisation and a loss of spontaneity.

As density of population increases, so areal specialisation results. The competition for space becomes so great that each area in the city tends to be put to the uses which yield the greatest economic return. Place of work becomes divorced from place of residence, and place and nature of work, income, habit, taste, preference and prejudice combine to produce a matrix of social worlds in the city. The close living together and working together of individuals who have no sentimental and emotional ties fosters a spirit of competition, aggrandisement and mutual exploitation. For those unable to find a secure life in some specialised role or sub-area the likelihood of dysfunctional behaviour increases, especially where densities are highest.

The heterogeneous nature of urban populations, Wirth believed, led to social instability and personal insecurity. Individuals acquired membership of widely divergent social groups each of which functions with respect to a single segment of his personality. Geographical and social mobility means that turnover in group membership is rapid so this further limits the possibility of establishing intimate and lasting relationships. As a consequence of a transitory habitat, there is little opportunity for the individual to obtain a conception of the city or to survey his place in the total scheme, and detachment from the organised bodies which integrate society leads to depersonalisation and isolation of the individual.

Wirth argued that changes of size, density and heterogeneity of human groups demand a response from the individual. The crowding of diverse types of people into a small area broke down existing social and cultural patterns and encouraged assimilation as well as acculturation — the melting-pot effect. He concluded that sooner or later the pressures engendered by the dominant social, economic and political institutions of the city would destroy primary group relationships and all the forces of social control which are derived from them. The end products would be 'anomie', a condition in which the normal rules and conventions which regulate social behaviour break down, 'alienation', in which the individual becomes detached from society, and social 'deviance'.

The Rural-Urban Continuum

In place of a polar distinction between urban and rural, Wirth introduced the idea of a progression of social and behavioural differentiation. At one extreme is the central area of the metropolis where, because of the large, dense and heterogeneous concentrations of population, urban ways of life are expected to be most clearly apparent; at the other, in the traditional folk society, rural patterns of interaction, association and behaviour will be found. Between these two ends of the spectrum, gradations of social and behavioural differentiation, reflecting the level of urbanism are postulated. Within three years of the publication of Wirth's paper, Redfield's study of the *Folk Culture of Yucatan* (1941) provided detailed empirical support for this contention. Redfield analysed in depth the social and economic characteristics of a tribal village (population 101), a peasant village (population 250), a town (population 1,200) and a city (population 100,000) which were held to be representative of the range of settlements to be found in Yucatan. His most important observation was that there were regular and pro-gressive variations among them, from the city of Merida at one extreme, to the isolated tribal settlement of Tusik at the other, and these differ-ences it was argued, were indicative of the existence of a continuous scale of rural-urban development:

> the peasant village as compared with the tribal village, the town as compared with the peasant village, or the city as compared with the town is (1) less isolated; (2) more heterogeneous; (3) characterised by a more complex division of labour; (4) has a more completely developed money economy; (5) has professional specialists who are more secular and less sacred; (6) has kinship and godparental insti-tutions that are less well organised and less effective in social control; (7) is correspondingly more dependent upon impersonally acting institutions of control; (8) is less religious, with respect to both beliefs and practices of Catholic origin as well as to those of Indian origin; (9) exhibits less tendency to regard sickness as resulting from a breach of moral or merely customary rule; (10) allows a greater freedom of action and choice to the individual (pp. 338-9).

Following Redfield's work, a wide range of studies have looked for evidence of these relationships among settlements along the rural-urban continuum, and it is with reference to this literature that the character-istics of life styles in modern urban society can be identified (Figure 4.1).

Considerable emphasis in life style studies has been placed upon the social and behavioural relationships of the residents of the central area of the metropolis. It is here that Wirth conceived the urban population

Figure 4.1: Selected Studies of Ways of Life Along the Rural-urban Continuum.

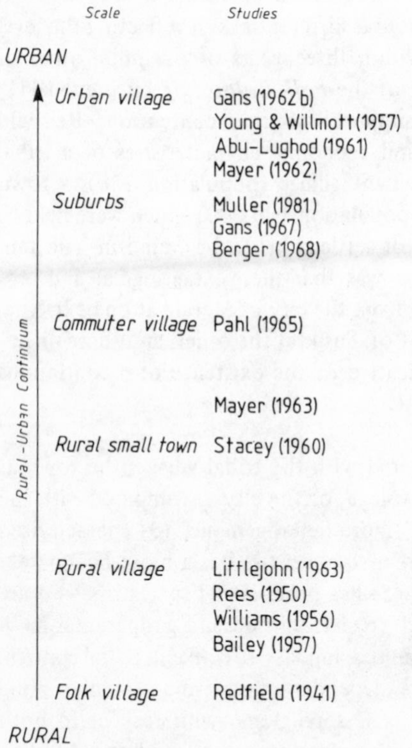

Scale	Studies
URBAN	
Urban village	Gans (1962 b)
	Young & Willmott (1957)
	Abu-Lughod (1961)
	Mayer (1962)
Suburbs	Muller (1981)
	Gans (1967)
	Berger (1968)
Commuter village	Pahl (1965)
	Mayer (1963)
Rural small town	Stacey (1960)
Rural village	Littlejohn (1963)
	Rees (1950)
	Williams (1956)
	Bailey (1957)
Folk village	Redfield (1941)
RURAL	

(vertical axis label: Rural - Urban Continuum)

as consisting of heterogeneous individuals, torn from past social systems, unable to develop new ones, and therefore prey to social anarchy, crime and deviance. Sociological surveys have, however, revealed the existence of several distinctive life styles in the area, some of which are characterised by a high level of social stability and conformity. In place of a single response determined by the size, density and heterogeneity of the social group, central areas offer a wide range of social and behavioural milieux.

For Gans (1962a) there are five basic types of inner city resident: the 'urban villagers', the 'cosmopolites' the 'unmarried or childless', the 'trapped and downward mobile' and the 'deprived'. The most well documented are the urban villagers, so called because they are members of small, intimate and often ethnic communities based upon inter-woven kinship networks and a high level of primary contact with familiar faces. The characteristics of urban villagers and their ways of life were described by Gans in his study of the residents of West End, Boston (1962b). The area was populated by immigrants from a variety of national and ethnic backgrounds including Italians, Poles, Irish, Greeks, Ukrainians, Albanians and Jews. Low incomes, an absence of occupational skills and qualifications and poor housing were common features. Behind the somewhat offensive façade of the area, which was strongly influenced by the dilapidated state of the buildings, the vacant lots and the garbage on the streets, Gans found a friendly, intimate and close-knit community reminiscent of that which exists in small towns and rural areas. Far from diminishing in importance, the family remained a major component in social organisation, and religion retained its hold on the people. The sharing of values was also encouraged by residential stability and the diverse network of personal acquaint-ances. Everyone might not know everyone else but as they did know something about everyone, the net effect was the same, especially within each ethnic group. Between groups, common residence, sharing of facilities and the constant struggle against absentee landlords created enough solidarity to maintain a friendly spirit. Although for many families, problems of unemployment, finance, illness, education and bereavement were never far away, neighbours and friends were always on hand to provide assistance and support. The relevance of Gans' generalisations may be criticised on the grounds that the West End was populated by predominently first generation immigrants who had yet to be exposed to the full impact of urban living, but this was not the case in Bethnal Green, London (Young and Willmott, 1957), Delhi (Bopegamage, 1957), Cairo (Abu-Lughod, 1961) and East London (Mayer, 1963), where similar urban village communities have been identified. Together, these studies point to the existence of a way of life in central areas which differs sharply from Wirth's urbanism. Far from reduced importance, life in the urban village is organised around kinship and primary groups which protect the individual from deviance and loss of personal identity.

Gans' second group, the 'cosmopolites' includes students, artists, writers, musicians and entertainers, as well as intellectuals and pro-fessionals who live in the city in order to be near the special 'cultural' facilities of the centre. Many are unmarried or childless, but others rear children in the city especially if they have the income to support the services of servants or governesses. Still others, though having an out-of-town residence, maintain a cosmopolitan life style from a *pied-à-terre* in the city. The 'unmarrieds or childless' are divided into two subtypes depending on the permanence or transience of their status. The temporarily unmarried or childless live in the inner city for only a limited time and upon marriage or starting a family, they leave for the outer city or suburbs, whereas the permanently unmarried live in the inner city for the remainder of their lives, their housing depending upon income. The former typically includes students or young people who share a downtown apartment or flat away from parents but close to jobs or entertainment opportunities. The fourth group are the 'trapped and downward mobiles' who are people who stay behind when a neighbourhood is invaded by non-residential land uses or by lower status immigrants, because they cannot afford to move, or are otherwise bound to their present location. Those who started life in a high class position but who have been forced down in the socio-economic hierarchy and in the quality of their accommodation because of personal circumstances are typical members of this group. It may also include old people, living out their existence on small pensions. The final group is the 'deprived' population; the very poor, the emotionally disturbed or otherwise handicapped, broken families and the non-white population. They must take the dilapidated housing and blighted neighbourhoods to which the housing market restricts them although among them are some for whom the slum is a refuge or temporary stop-over to save money for a house in the outer city. These five types all live in dense and heterogeneous surroundings in the central area yet they have such diverse characteristics and ways of life that it is hard to see how density and heterogeneity could exert a common controlling influence. The 'cosmopolites' and the 'unmarried and childless' live in the inner city because they wish to, the 'urban villagers' are there partly because of necessity, partly because of tradition. The final two types are in the inner city because they have no option. For Gans, life styles in the inner city involve some element of choice which in turn is heavily conditioned by background and class; they are largely independent of location.

A second major focus of study within the rural-urban continuum is concerned with the life styles in the suburbs. As an area of the city, the suburbs date from the inter-war period in the UK and from the immediate post-war era in North America when large scale residential development at what was then the edge of the city took place. They were largely a response to the decentralisation of the urban population which, though dating back to the late nineteenth century, was made possible by extensions of suburban rail networks and increases in private car ownership. Suburban growth was especially fast in many US cities in the decade following 1945 when post-war prosperity, combined with a rapid increase in the rate of household formation, generated enormous demands for housing. The most important feature of residential construction was that comparatively few designs were employed so that the suburbs are characterised by streets or avenues of properties of broadly similar size and style. Today 40 per cent (84 million) of the US population lives in the suburbs as compared to 30 per cent in central cities and 30 per cent in non-metropolitan rural areas. The suburbs, moreover, continue to be the most rapidly growing of these three categories of residential area (Muller, 1981, p. 4).

Life styles in the suburbs were first subjected to detailed analysis in the early 1960s (Gans, 1967). The suburbs at that time were in a comparatively early stage of development and consisted predominantly of young, married, child-rearing, middle income, white collar groups. They were characterised by single family rather than by multi-family occupation and were therefore more socially homogeneous than other parts of the city. Suburbanites were observed to be upwardly mobile because they were predominantly young, and because they were employed as engineers, middle managers, lawyers, salesmen, insurance agents, teachers and bureaucrats in expanding professions. Most of these jobs required training and qualifications so that suburbanites were also distinguished by their above average levels of education.

According to their position along the rural-urban continuum, the suburban population should exhibit deviant, anomic and alienated life styles that are akin to, but not as extreme as, those held to exist in central areas. Size, density and heterogeneity of population are far greater in the suburbs than in rural areas so that secondary relationships will prevail, and suburban society will be typified by impersonality, isolation and disaffection. Studies of suburban life styles undertaken twenty years ago, however, provided little support for these expectations. They identified what Gans (1962a) termed a 'quasi-primary' way of life based upon close but guarded patterns of relationship with others. The suburbs were seen as an area with an active social and community life, characterised by informal visiting and participation in clubs and associations. There was little anonymity or privacy. Involvement extended beyond the limits of voluntary associations to include an equally active participation in local civic affairs. Solidarity permitted strong if informal sanctions to be applied to those who failed to conform in terms of upkeep of housing, personal or social behaviour. The strength of quasi-primary relationships was indeed such that they lead to a general sense of social and areal identity so that the suburbs could be divided into a number of distinctive neighbourhoods, each with their distinctive life styles and sense of place.

If life styles in the immediate post-war suburbs owed little to Wirth's urbanism, this is even more the case today. A feature of urban change in the US in the last decade has been the increasing suburbanisation of lower income groups, non-professionals and blacks so that the initial homogeneity of the suburbs has been destroyed. For Muller (1981, p. 70), following Suttles (1975, pp. 265-71) and Berry (1973, p. 65) there are four basic community forms within contemporary American suburbia: the exclusive/affluent apartment complex; the middle class family areas; low income ethnic centred working class communities; and cosmopolitan centres. The socio-economic characteristics of these types in terms of income, population stability, age structure and education are illustrated by data for representative communities in Grosse Pointe Shores, Michigan; Darien, Connecticut; Levittown, New York; Milpitas, California; and Princeton, New Jersey (Table 4.2).

The high income suburbs are characteristically located in areas possessing both physical isolation and the choicest environmental amenities around water, trees and higher ground. Since houses are built on large plots and are well fenced, neighbouring is difficult, and people keep in touch by participation in local social networks. The latter are tightly structured around organisations such as churches, country and golf clubs, and newcomers to the community are carefully screened for their social credentials before being accepted. Exclusiveness is reinforced by private schools and by the emphasis placed upon class traditions. A recent development in high income areas is the growth of luxury apartment and condominium complexes which attract increasing numbers of affluent singles, families and senior citizens.

Middle class family suburbs are located as close as possible to the high status residential enclaves of the most affluent. They are populated by middle income groups which are arranged into nuclear family units. The management of children is a central concern and most local social contact occurs through family oriented formal organisations such as school associations and children's societies and sports clubs. Despite the closer spacing of homes and these integrating activities, middle class suburban-ites are not communally cohesive to any great degree. Emphasis upon family privacy, and freedom to pursue upward social mobility aggressively does not encourage the development of extensive local social ties.

Table 4.2: Selected Socio-economic Characteristics of Representative Suburban Life Style Community Types, 1970.

Variable	Exclusive Upper Income Grosse Pointe Shores, Mich.	Middle-income Family			Cosmopolitan Suburb Princeton, NJ
		Upper Darien, Conn.	Lower Levittown, NY	Working Class Milpitas, Calif.	
Median family income	$32,565	$22,172	$13,083	$11,543	$12,182
Per cent families > $25,000	60.4	41.9	5.7	1.7	18.0
Per cent families < $15,000	19.1	29.6	64.0	75.8	59.8
Per cent families in poverty	2.1	2.4	3.0	5.3	5.2
Per cent black	0.2	0.5	0.1	5.2	10.0
Median age	41.3	32.1	24.0	20.9	28.4
Per cent population < 18	30.3	35.5	41.4	46.4	16.3
Per cent population > 65	12.4	7.8	3.6	2.1	10.6
Per cent same address 1965	59.5	60.3	75.1	31.5	34.2
Per cent population 3-34 in school	80.3	71.0	63.0	55.3	67.4
Per cent high school graduates	83.1	83.1	64.2	63.6	76.5
Per cent women in labour force	20.8	34.5	41.9	45.5	49.4
Per cent professional and technical workers	28.3	25.7	13.6	16.9	37.6
Per cent managerial/administrative workers	31.6	25.2	9.4	5.2	7.5
Per cent operatives	0.8	4.3	7.6	18.5	2.8

Source: Muller (1981), p. 79.

Neighbouring (mostly child related) is limited and selective, and even socialising with relatives is infrequent. Most social interaction revolves around a non-local network of self-selected friends, widely distributed in suburban space. The insular single family house and dependence on the automobile for all trip-making accommodates these preferences and fosters a congruence between life style and the spatial arrangement of the residential environment.

Outside the central city, the working class and poor areas are to be found in the innermost pre-automobile suburbs, and adjacent to industrial areas and railroads. For Muller (p. 71), working class suburban life styles differ from middle class suburban life styles in a number of ways. Whereas middle class suburbs stress the nuclear family unit, socialisation with friends and a minimum of local contact, working class neighbour-hoods accentuate the extended family, frequent home entertaining of relatives rather than acquaintances and a great deal of informal local social interaction outside the home. The latter is reflected by the importance of local meeting place such as churches, taverns and street corners. Rather than a breakdown of social contact, these informal networks introduce an important element of social cohesion into working class suburbs. Moreover, local area attachment is reinforced by a person oriented rather than a material or achievement oriented outlook. Working class suburbanites have no great hopes of getting ahead with their largely blue collar jobs, have few aspirations as regards upward social mobility and, therefore, view their present home and community as a place of permanent settlement.

Muller's fourth category of suburban life style is the suburban cos-mopolitan centre. It is distinguished by communities of professionals, intellectuals, students, artists and writers who participate in far-flung intra- and inter-metropolitan social networks and communities of interest. As theatres, music and arts facilities, fine restaurants and other cultural activities have deconcentrated, so cosmopolitan centres have spread throughout the city, and life styles that were once the preserve of inner areas have become predominantly suburban. This expansion has been assisted by the opening of branch campuses of universities and colleges which provide a cultural and intellectual focus in the suburbs.

The very different types of community and associated life styles that can be identified in the contemporary American suburbs suggests that location is unimportant as a controlling factor. What differentiates suburban communities is occupation, social class and ethnicity and these are largely independent of size, density and heterogeneity considerations. Especially important are considerations of income which determine both residential location and patterns of social interaction. The suburbs, like the central city, offer a range of distinctive niches within a mosaic culture which is increasingly dominating urban America. Life styles are not determined by position within a rural-urban continuum, rather they reflect the socio-economic characteristics of the population.

A third group of 'urban' life style studies is concerned with the urbanised fringes of 'rural' areas around large cities. In view of the intermediate position along the continuum, the population of commuter zones may be expected to exhibit features of both an urban and rural way of life. Pahl's (1965) study of the norther commuter belt of London identified the major groupings in the metropolitan village and explored their life styles. They include large property owners, the salariat, retired workers with some capital, urban workers with limited capital or income, rural working class commuters and traditional ruralites.

The first group is the large property owners. They are tied closely to the village by tradition and land holdings but may have considerable financial and business connections elsewhere. Amounting perhaps to only one or two families in each village, they may be absent for long periods and so play no real part in daily village life. Some indeed may be internationally oriented. The professional and managerial salariat choose to live in a commuter village not merely for the physical surroundings, but also on account of the distinctive pattern of social relationships which they associate with rural communities. They are attracted by the possibility of interaction with other high status groups and by the perceived 'friendliness' of village residents, although the degree of mixing, especially with manual workers and their families, is limited. The salariat and the manual worker commonly send their children to different schools, perform different roles in local village associations, and hold contrasting opinions over local issues such as conservation and rural traditions. Retired urban workers with some capital are a third group who choose to live in a village because it is perceived to offer an attractive locality in which to spend old age. Because of age, family ties and financial independence, they participate in a distinctive pattern of social interaction much of it non-village based.

Pahl's fourth category is that of urban workers with limited capital and income and includes those who are compelled on financial grounds to seek cheap housing in the metropolitan village. Although reluctant commuters, they depend heavily upon local services and so integrate comparatively readily into the village community. In contrast to this group, the rural working class commuters are traditional village residents who find that their work is outside the village. The nature of employment centralisation in many rural areas in recent years suggests that this is an expanding group. The final category, that of traditional ruralites, comprises a small minority element of local tradesmen and agricultural and related workers whose residence and employement are both local. There may be close kinship and other ties with the rural working class commuter and in practise it is difficult to distinguish them in sociological terms.

Despite a common location, Pahl's analysis shows that life styles in the metropolitan village are highly differentiated. As in the inner city and the suburbs, size, density and heterogeneity of population do not impose a common social or behavioural response. 'There are some people who are in the city but are not of it (the urban villagers), whereas others are of the city but are not in it (the mobile middle class of the metropolitan commuter village)' (Pahl, 1968, p. 273). Rather than being simply locationally determined, life styles appear instead to be a function of constrained individual choice.

Studies of life styles in central areas, suburbs and metropolitan villages appear to provide little support for Wirth's ideas. It is important, however, to put both the theory and the empirical evidence into a clear historical perspective. Most of the studies which have been reviewed here were published in the last two decades, twenty years after the appearance of *Urbanism as a Way of Life,* which itself was based upon observations of earlier Chicago ecologists, some of which referred to the late nineteenth century city. The pace and nature of subsequent urban development suggests, however, that contemporary urbanism exhibits many characteristics and properties that were not present in urban life eighty years ago. Wirth was writing, as Gans (1962a) has observed 'during a time of immigrant acculturation and at the end of a serious depression, an era of minimal choice. Today it is apparent that high density heterogeneous surroundings are for most people, a temporary place of residence . . . they are the result of necessity rather than choice. As soon as they can afford to do so, most Americans head for the single family house and the quasi-primary way of life of the low density neighborhood, in the outer city or suburbs.' For Berry (1973, p. 36), this change of social context means that Wirth's work has not so much been refuted as been overtaken by events. 'Wirth believed that he offered a theory upon which to build future research. In actuality, his theory was a peroration (concluding statement) on a city that had passed.'

Alternative Theories of Urbanism

In considering the factors which determine life styles in the contemporary city, it seems clear that social class, ethnicity and stage in life cycle are of particular importance. These characteristics influence both an individual's social and locational aspirations, and also the extent to which he will achieve those goals. As such they form the basis of an alternative, compositionalist theory of urbanism. This theory interprets and accounts for different life styles in terms of an individual's personal circumstances rather than as a mass response to urban living.

In contrast to the determinist approach as represented by Wirth, the compositionalists do not believe that city living weakens small primary groups, rather they maintain that these groups carry on undiminished (Table 4.3). In place of size, density and heterogeneity, individuals' behaviour is held to be the product of social class, ethnicity and stage in life cycle. These characteristics determine patterns of interaction and association and so give rise to a mosaic of social worlds based upon kinship, neighbourhood, occupation, education or similar personal attributes. They are exemplified by immigrant enclaves such as Little Italy (New York) or Notting Hill (London), by upper class colonies including Nob Hill (Boston) and the Gold Coast (Chicago) and by bohemian quarters such as the West Bank (Paris) and Greenwich Village (New York); but these are only the most obvious and well known of the many types of social worlds that are a product of the urban mosaic. Rather than be destroyed, these private milieux endure and flourish in even the most urban of environments.

Table 4.3: Alternative Explanations of Urban Life Styles

	Leading Proponents	Key Variables	Effects of Urban Life on Social Groups	Socio-psychological Consequences
Determinist	Wirth (1938)	Size, density and heterogeneity	Breakdown of primary groups	Alienation, deviance anomie
Compositional	Gans (1962a, 1962b, 1967); Lewis (1952); Pahl (1970)	Class, ethnicity, stage in life cycle	No direct effect	No direct effect. (Indirect effects as a consequence of class, ethnic and life style position)
Subcultural	Fischer (1976)	Size, 'critical mass', interaction	Creation of primary groups	Subcultural integration

Compositionalists do not suggest that living in cities has no socio-psychological consequences, but they do maintain that any *direct* effects are insignificant. If community size does not have any consequences, they result from the ways in which it affects the position of individuals in the economic structure, the ethnic hierarchy and the life cycle. For example, large communities may provide better paying jobs, and the people who obtain them will be deeply affected but they will be affected by the new economic circumstance and not by the urban experience itself. Similarly, the city may attract a disproportionate number of males so that many of them cannot find wives. This will certainly affect their behaviour but not because the city has cut their social ties. Fischer (1976, p. 35) explains the contrast between the determinist and the compositional approaches in this way: 'both emphasise the import-ance of social worlds in forming the experiences and behaviour of individuals but they disagree sharply on the relationship of urbanism to the viability of those personal milieus. Determinist theory maintains that urbanism has a direct impact on the coherence of such groups with serious consequences for individuals. Compositional theory maintains that these social worlds are largely impervious to ecological factors and that urbanism thus has no serious *direct* effects on group or individuals.'

The contrasting viewpoints represented by the determinist and the compositional approaches have been synthesised by Fischer (1976) in the form of a subculture theory of urban life styles. Subculture theory contends that living in cities independently affects social life not by destroying social groups as Wirth's deterministic approach suggests, or by leaving them untouched as the compositionalists believe, but instead by helping to create and strengthen them. The most significant social consequence of community size is the promotion of diverse subcultures (culturally distinct groups such as college students or West Indian immigrants). Like compositional theory, subculture theory maintains that intimate social circles persist in the urban environment. But like determinism it maintains that ecological factors produce significant effects on the social ordering of communities precisely by supporting the emergence and vitality of distinctive subcultures.

Fischer's subculture theory holds that there are two ways in which urban mosaics are produced. The first is that large communities attract immigrants from a wider area than do small towns and so they receive a great variety of cultural backgrounds which contribute to the formation of a diverse set of social worlds. The second is that as predicted by the determinists, large size produces a differentiation of occupational and social functions. Where the population size is large enough, what would otherwise be only small groups of individuals becomes a vital, active subculture. Sufficient numbers allow them to support institutions such as clubs, newspapers and social functions which serve the group and allow them to have a visible and coherent identity. For example, if only one person in a thousand is intensely interested in amateur dramatics, in a town of only 5,000 there would only be five people, enough to do little else but engage in conversation about acting. But in a city of one million, the thousand interested individuals would be sufficient to support plays, to lay on special visits to the theatre; and to maintain a special social milieu. Interaction between members of a social world and those outside it may serve to break down barriers, but a more common reaction is for a defensive retreat into one's social world so that the cohesion of the group is intensified and reinforced. Subcultures do not therefore exist because social worlds in the city break down under the impact of ecological forces, but rather because within large, dense and heterogeneous populations their formation is facilitated and encouraged.

Urbanism in an Aspatial World

The deterministic, compositional and subcultural approaches provide alternative explanations of the different patterns of association, inter-action and behaviour that may be experienced in cities. An important characteristic of contemporary life styles is, however, that they are becoming ever more independent of a city, indeed of any location. Although individuals live in a particular place and participate in com-munity life in and around that place, they are increasingly able to maintain contacts with others on an interest basis, over progressively greater distances and so become members of interest communities which are not territorially defined. With modern transportation and telecommunications services, a choice of life styles, once the prerogative of those actually resident in and around the city, is open to all.

For Webber (1964), the extent and range of participation in interest communities is a function of an individual's specialisation. The more highly skilled a person is, or the more uncommon the information he holds, the more spatially dispersed are the members of his interest group and the greater the distances over which he interacts with others. There is an hierarchical continuum in which the most highly specialised people are participants in interest communities that span the entire world; others, who are less specialised seldom communicate with people outside the nation but regularly interact with people in various parts of the country; others communicate almost exclusively with their neigh-bours (Figure 4.2). For example, the research worker will be a member

Figure 4.2: Webber's Concept of the Structure of Interest Community by Realm.

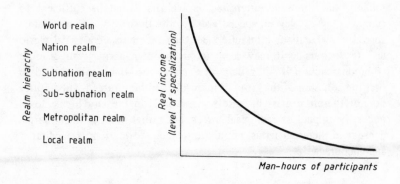

Source: Webber (1964), p. 125.

of a wide range of interest communities and will spend more of his time participating in national and international scale interest communities than will the primary school teacher. Similarly, the upper limit of the interest community of the school caretaker will in all probability be restricted to the city. All three will, however, be involved in more parochial interest groups in association with their roles as parents, amateur sportsmen, or members of local clubs and societies.

For any given level of specialisation, Webber argued that there are a wide variety of interest communities whose members conduct their affairs within roughly the same spatial field and which he refers to as urban realms. Urban realms are neither urban settlements nor territories, but heterogeneous groups of people communicating with each other through space. They are somewhat analgous to urban regions, but, contrasting with the vertical divisions of territory that are organised around cities and accord with the place conception of regions, urban space is divided horizontally into a hierarchy of non-place urban realms (Figure 4.3). Irrespective of location, every person is at different moments a communicant in a number of different realms as he shifts from one role to another, but only the most specialised people communicate across the entire nation and beyond. In this context the city is no longer a unitary place. Rather it is a part of a whole array of

Figure 4.3: Webber's Concept of Role Specialisation and its Geographical Relationships.

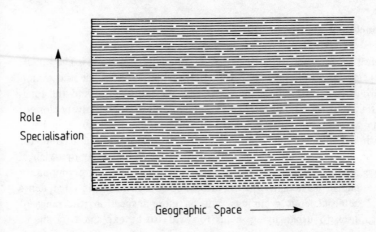

Source: Webber (1964), p. 119.

shifting and interpenetrating realm spaces.

The paradox with modern telecommunications media is that in a sense they are re-introducing the close inter-personal contacts and informal mechanisms of control that are reminiscent of village life. For McLuhan (1962, 1964) history is made up of three stages. The first is the pre-literate or tribal stage in which people live close together and communicate orally. This is followed by the Gutenberg or individual stage in which communication takes place by the printed word, and thinking is done in a linear-sequential pattern. The third is the neo-tribal or electric stage in which computers, television and other electronic communication media have moved men back together. Urbanisation as observed by Wirth and as reflected in the emergence of different life styles is most closely associated with the individual stage. It is at this time that the predominant use of written communication necessitated the adoption of a strict hierarchy of responsibility and a clear separation of roles and functions. With electronic communication, social distances are reduced, indeed removed, and specialisation of knowledge is replaced by generalisation. This intimate electronically linked community has, however, little in common with the traditional rural way of life. Rather, it amounts to a global village, a world characterised by a homogeneous urban culture in which we may all be participants.

Conclusion

Life styles in modern society may take a number of forms. The values, expectations and aspirations of the farm labourer have little in common with those of the bohemian, the affluent suburbanite and the long-distance urban commuter, and yet all may reside comparatively close to one another, in and around the city. Theories of urbanisation begin with traditional rural life styles and seek to identify and account for their progressive replacement by urban ways of living. At first glance, it seems reasonable to associate such changes with increasing community size, and in particular with the effects of concentration in large, dense and heterogeneous groupings, but on close inspection, the range of responses to urban living seems too wide to be explained by simple locational determinism. What modern urban society offers is a mosaic of social worlds characterised by class, ethnic and age differences. The extent to which individuals are free to choose among these life styles is clearly dependent upon their own personal circumstances. It is widest in the case of high income, professionally qualified mobile groups, but is non-existent for members of the poverty-stricken underclass. With modern telecommunications services, however, and the rapid trans-mission and diffusion of attitudes, values and expectations that they imply, participation is becoming increasingly independent of location. Life styles which were once the preserve of those actually resident in the city are becoming equally available to all.

Urbanisation is social change on a vast scale. It means deep and irrevocable transformations that affect every aspect of social life and all sections of society. There is little doubt that such changes were initiated by the explosive growth of large cities that began in the late eighteenth century. Equally, it seems clear that in advanced Western societies its effects and ramifications are now all pervading. What was once a simple rural-urban gradient is now a continuous urban realm. This fusion of rural and urban has, however, yet to begin in the many countries of the underdeveloped world in which the population remains predominantly village based and remote from urban influences. Rural ways of life of an historic and a traditional nature continue unchanged for perhaps a third of the world's population. Despite its declining relevance to the advanced nations, the study of urbanisation will continue though in a different cultural context. Some of the most important questions for the next decade surround the speed, characteristics and consequences of the adoption of Western urban life styles by rural Third World populations.

5   URBAN LOCATION AND THE URBAN SYSTEM

It is clear from even the most casual observation, that the modern city performs a variety of functions. Shops, warehouses and theatres testify to a servicing role, workshops and factories highlight the importance of manufacturing and assembly, while office blocks provide visual evidence of administrative and managerial responsibilities. Though present in most cities today, these urban functions developed at different times in the past and vary in their contemporary importance. For example, the pre-industrial economy was oriented towards serving the needs of agriculture so the preferred locations for settlement were those places which combined security and accessibility to the widest markets. Similarly, manufacturing developed on sites which presented the best opportunities for minimum cost production, and where these coincided with existing pre-industrial settlement, the city developed as both a service and an industrial centre. Management has emerged only recently as a major source of employment in business and government and in expressing a preference for the central areas of large cities gives those places an even wider occupational profile. These characteristics and requirements mean that there can be no general explanation of urban location. Rather, there are a set of approaches, each concerned with the location of individual urban functions.

Traditionally, geographers sought to explain the distribution of urban settlements in terms of an analysis of site and situation, an approach which produced elaborate classifications of positions, but little understanding of the principles of location involved. Today, emphasis is placed upon modelling those social and economic relationships and processes which determine the urban geographical pattern (Table 5.1).

Table 5.1: Urban Location: Major Theoretical Contributions

Cities as service centres	Christaller (1933)
Cities as manufacturing centres	Weber (1909), Hoover (1948), Lösch (1954), Greenhut (1956), Isard (1956)
Cities as management centres	Thorngren (1970), Tornqvist (1970), Warneryd (1968), Goddard (1973)

The model building approach had its origins in the classical location theory of Thunen (1826) and Weber (1909), and is represented by the more recent statements of Hoover (1948), Lösch (1954), Greenhut (1956) and Isard (1956) which are concerned with the distribution of manufacturing activity. It received a major stimulus from Christaller's *Central Places in Southern Germany* (published in German in 1933 and available in English translation in 1966), which contained a comprehensive statement on the size, spacing and functions of service centres. Models of the location of management occupations stress the fundamental importance of business communications and information availability as the work of Warneryd (1968), Thorngren (1970), Tornqvist (1970) and Goddard (1973) has shown. This chapter examines the factors which determine the spatial distribution of settlements by reference to models of the location of service, manufacturing and managerial activities and their organisation as a spatial system. Emphasis is initially placed upon the central place model as a theory of the location of market centres in simple agricultural economies.

Cities as Service Centres

The classical central place model, which consists of Christaller's statement, together with refinements and embellishments proposed by Berry and Garrison (1958a, b), provides a theory of the size, function and spacing of market centres. It begins with a series of assumptions designed to simplify the characteristics of the marketing system:

(1) An unbounded plain with uniform resource endowment;
(2) an even distribution of population and purchasing power;
(3) equal freedom of movement in all directions;
(4) transport costs are proportional to distance travelled; and
(5) goods and services which have the same basic price at any centre where they are offered for sale.

Together, these assumptions define the homogeneous surface known to classical economists as an isotropic transport plain. Its overriding characteristic is that no location enjoys any advantage over any other location in terms of ease or cost of access, or the price of goods and services which are sold there.

Within this landscape, limits are imposed upon geographical and economic behaviour. These are:

(1) supply and demand must be equated as far as possible;
(2) the sum of distances travelled must be as small as possible;
(3) profits must be maximised; and
(4) the total number of centres serving the area should be as small as possible.

These behavioural constraints produce in effect a capitalist marketing system inhabited by 'Economic Man', that mythical individual who seeks to minimise costs and maximise returns by making wholly rational decisions on the basis of full and comprehensive knowledge of all aspects of the market.

The environmental assumptions and behavioural constraints create an idealised world in which most of the major factors which affect the location decision process are regulated and controlled. This procedure isolates the spatial variable for detailed independent consideration. The effects of distance are introduced in the form of transport costs. For any consumer, the cost of any commodity involves the actual market price of goods and services, which is the same for everyone, plus the expenses incurred in the return journey from home to place of purchase. Prices increase uniformly with distance (in all directions) away from any point of sale, and, as prices rise, there is a corresponding reduction in demand (Figure 5.1). At some distance, a point will be reached at which the price will become so high as to exclude purchase. This point of nil demand is known as the maximum range and defines the outer limit of sales for any commodity or service. In areal terms this means that each point of sale will be surrounded by rings which delimit the maximum market area for any item or service sold.

An inner sales area, known as the threshold market area, is determined by the economics of supply. Each retailer who wishes to enter into business must be satisfied that he will be able to sell sufficient items of stock to cover basic operating costs such as rates, rents, mortgage repayments, salaries, depreciation, etc. and bring in a worthwhile profit. Without this minimum volume of sales, outgoings exceed returns and the enterprise fails. The assumption of uniformly distributed purchasing power gives this commercial principle a spatial meaning: each centre is surrounded by a ring which encloses a sufficient number of consumers to generate this minimum turnover. Sales made to people living outside the threshold market area, but within the maximum range area, yield 'excess profits' to the retailer (Figure 5.1).

The maximum number of suppliers who can operate profitably in the market is determined by the threshold values of goods and services.

Figure 5.1: Classical Central Place Theory: Price, Distance and Demand Relationships in the Isotropic Transport Plain.

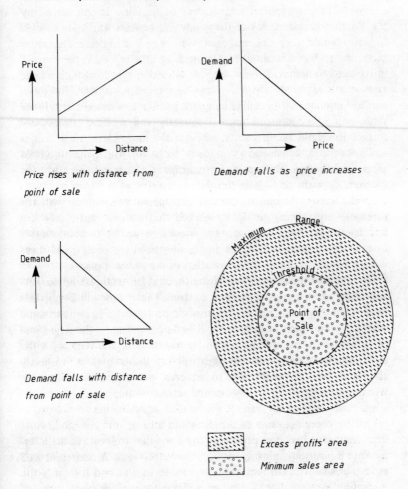

Price rises with distance from point of sale

Demand falls as price increases

Demand falls with distance from point of sale

With a threshold of 100 units of demand a week, and a total market of 10,000 units per week, a maximum of 100 suppliers will be able to operate. The geographical arrangement of suppliers is determined by the thresholds and ranges of the goods and services which are offered for sale. Consider a number of commodities ranked in descending order according to their threshold requirements. At the top are expensive and infrequently purchased items such as television sets for which a large sales area is required before they can be retailed profitably. The intermediate rankings are occupied by goods and services which are of lower value and are consumed more frequently and so have successively smaller threshold requirements, while at the lowest level are groceries and foodstuffs, consumed on a daily basis, for which the threshold areas are minimal. The optimal arrangement of centres to supply the commodity with the highest threshold requirement is shown in Figure 5.2. The system is one of competitive spatial equilibrium with the threshold areas being so tightly packed that no one centre enjoys a locational advantage. Even so, there are excess profits to be made out of sales to consumers living in the gap between the circles, and these will be divided equally between retailers in the adjacent places.

Suppliers of the commodity having the next largest threshold markets will also locate in the existing A centres. There they will cover their basic operating costs, make a worthwhile profit and also share in some excess profits which in this case will be larger owing to the larger area of interstice. They will be joined by suppliers of the item with the third largest threshold market who will enjoy even greater prosperity since they will share in sales to an even greater excess profit area. Working down the array of goods and services in this manner, however, a marginal point at, say, rank 8 will be reached where the area of excess profits becomes the same as the threshold area of that commodity. At this juncture it becomes profitable for a supplier to locate in the intermediate B positions, midway between the surrounding A centres, as well as in the A centres themselves. This introduces a second layer into the marketing system. The A centres will supply the complete range of goods and services, but B centres offer only those items with thresholds below that of the first hierarchical marginal good.

The same line of argument can now be applied to any one of the A centres and any two of the B centres. Again, working down the array of goods, suppliers of the commodity with the next largest threshold to that of the first hierarchical marginal good will locate in the existing centres, as will suppliers of the items with the next largest threshold, and so on. At some stage, a second marginal point will be reached where

Figure 5.2: Classical Central Place Theory: the Derivation of the Functional Hierarchy of Service Provision.

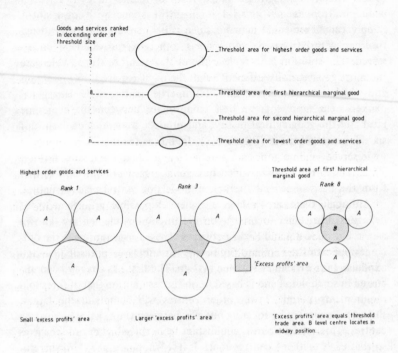

the area of excess profits will exactly equal the threshold area of a commodity. When this occurs, suppliers will again find it profitable to locate midway between the A and B centres at C, as well as in the A and B centres themselves, so adding a third tier to the system. Retailers operating at this level sell only those goods and services which have a threshold requirement which is the same as or smaller than that of the second hierarchical marginal commodity. They supply a more limited range of goods and services to a smaller market than do the A and B level suppliers, therefore reinforcing the hierarchical arrangement of centres and market areas within the network. These arguments are repeated, so adding further distribution levels to the marketing system, until the array of goods is exhausted.

Having proposed a pattern of goods and service provision, central place theory considers the spatial arrangement of market areas. The arguments are essentially geometrical and relate to the packing properties of the circle. Although the radial form of market areas is a logical deduction from assumptions 1-4, it remains at variance with constraint 1. Supply can never equal demand since radial markets cannot be fitted together without leaving some areas either overprovided or under-served. The solution is to replace the circles with hexagons, which are the most geometrically efficient of all the regular space filling polygons and which most closely retain the properties of the circle. Hexagonal market areas represent the best compromise between the economic ideal and the geographical reality, and produce a hierarchical lattice of six-sided market areas (Figure 5.3). In this network, no consumer is underserved, and no goods are purchased at an unacceptable total price. Variations in the ways in which hexagonal areas can be fitted together form the basis of several alternative models of central place structure.

Although there are a large number of possible arrangements of hexagons, Christaller focused upon the three most elementary, known as the K = 3, K = 4 and K = 7 networks. These forms represent the outcomes of attempts to maximise the distributive, accessibility and administrative efficiencies of the system. The first has already been discussed in some detail and is based upon the assumption that, if threshold requirements permit, lower order centres will command the largest share of the market by locating midway between the existing competing centres. This vertex point is equidistant from three higher order centres so that each of the A centres controls its own, plus a one third share of the six surrounding B level market areas. Inspection of Figure 5.3 shows that this rule of threes progression for market areas holds for the hierarchy as follows:

$$A : B : C : D : E : F = 1 : 3 : 9 : 27 : 81 : 243$$

while the number of centres present in successive levels is:

$$A : B : C : D : E : F = 1 : 2 : 6 : 18 : 54 : 162.$$

The K = 3 network provides the optimum arrangement of centres for marketing purposes but is costly in terms of transport provision. Location of places at the vertices of the hexagons ensures the maximum possible spacing of settlements and means that lower order centres cannot be aligned with higher order centres so as to minimise overall accessibility.

Figure 5.3: Classical Central Place Theory: the Structure of the K = 3 Network.

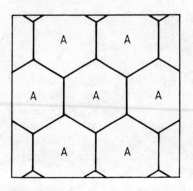
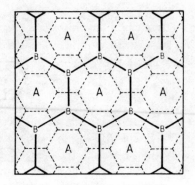

K = 3

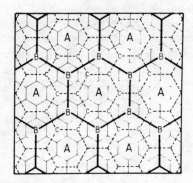

These difficulties can be overcome by reorganising the system according to a 'transport principle'. Once the basic distribution of the A centres is derived, the B centres are located halfway between two of them so that they are positioned at the mid points of the sides, rather than at the vertices of the hexagonal market areas (Figure 5.4). Each A centre serves its own market area and has a half share in the sales of highest order goods to each of the six surrounding centres so that the progression of market areas is:

Figure 5.4: Classical Central Place Theory: the K = 3, K = 4 and K = 7 Networks.

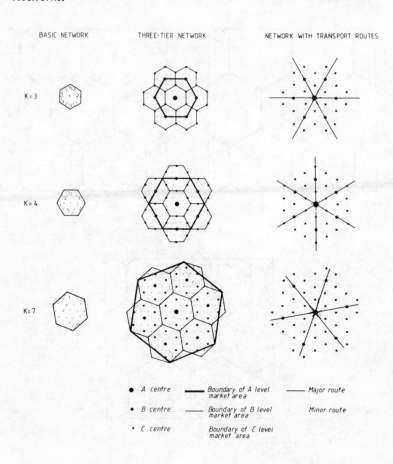

A : B : C : D : E : F = 1 : 4 : 16 : 64 : 256 : 1024

and the number of centres is:

A : B : C : D : E : F = 1 : 3 : 12 : 48 : 172 : 688.

The transport network which serves this system is regarded as optimum.

The major centres are both closely spaced and regularly aligned so that the total route mileage is minimised. The network consists of an alternating series of trunks, major and minor roads which link the highest, middle and lowest orders of settlement respectively.

Christaller's K = 7 network is based upon the 'separation principle' and provides the most efficient system for administrative purposes. A feature of both K = 3 and K = 4 is the borderline location of lower order centres which means they are 'shared' by the adjacent higher order places. This division of economic allegiance is hardly commensurate with efficient government and administration, so in the K = 7 network the market boundaries are made to circumscribe rather than to divide centres. This produces a far more efficient areal division of power in which each higher order centre completely controls a surrounding ring of six lower order places. The K = 7 progression of market areas is:

$$A : B : C : D : E : F = 1 : 7 : 49 : 343 : 2,401 : 16,807$$

while the number of centres present in each level is:

$$A : B : C : D : E : F = 1 : 6 : 42 : 294 : 2,058 : 14,406.$$

In this rather complex arrangement, B level centres are directly linked to the A level metropolitan centre, but as a consequence of the very large number of lower order places in the system, the network is characterised by many minor routeways (Figure 5.6).

These three K networks integrate locational activities and spatial linkages in a complex theoretical statement on settlement structure. Based upon a series of simplifying assumptions and constraints, classical central place theory predicts the optimal organisational and spatial arrangement of centres for administrative, marketing and transport purposes. From this interrelated set of propositions, there are two logically deducible hypotheses. The first is that there will be a functional hierarchy of service centres, consisting of tiers of settlements, each containing places which offer similar types of goods and services to similar sized market areas. The second is that urban settlements will be uniformly spaced.

Empirical research in a wide variety of locations has been undertaken to assess the validity of classical central place theory (Table 5.2). Studies of the hierarchy of service centres involve an examination of the range of shops and retail outlets present in each place, the size of their market areas, and the extent to which their functional complexity

Table 5.2: Classical Central Place Theory: Selected Empirical Studies.

Country	Area	Study
Germany	Southern Germany	Christaller (1933)
	Baden-Wurttemberg	Barnum (1966)
USA	Wisconsin	Brush (1953)
	Washington State	Berry and Garrison (1958a)
	Iowa	Berry (1967); Golledge *et al.* (1966);
	Twenty study areas	King (1962)
UK	Wales	Rowley (1970, 1971)
	Mid-Wales	Lewis (1970)
	Rhondda Valleys, Wales	Davies (1967, 1968, 1970)
	Yorkshire	Tarrant (1968)
	East Anglia	Everson and Fitzgerald (1969)
Yugoslavia	Yugoslavia	Vriser (1971)
Australia	Tasmania	Scott (1964)
	Australia	Johnston (1964)
New Zealand	Auckland	Badcock (1970)
India	Punjab	Mayfield (1967)
Canada	Ontario	Marshall (1969)
		Preston (1971, 1979)

correlates with size. The research involved is illustrated by Berry's (1967) study of central places in southwest Iowa (Figure 5.5). A five level hierarchy was identified in the area consisting of a regional capital (Council Bluffs-Omaha), cities (which include Atlantic, Corning, Red Oak and Glenwood), towns, villages and hamlets. Despite the wide range of centres, patterns of consumer movement are distinctive and largely non-overlapping, and define trade areas for different services. As predicted by the theory, rural residents patronise all central place levels for lower order goods such as groceries, but for the purchase of women's coats and dresses, and for visits to a lawyer and a physician, distances travelled are greater, and services are provided at fewer centres (Figure 5.6). For more specialised services, such as hospital care, four centres — Council Bluffs-Omaha, Atlantic, Red Oak and Harlan — dominate the area. An important feature of the geography of service provision is that the trade areas of small centres nest within those of the larger centres so that overall consumer movements are as short as possible. Sales of newspapers in fact show that for highest order servicing, parts of the area look to the east, to the larger and more distant Des Moines.

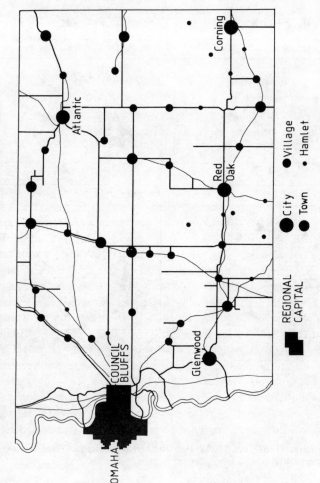

Figure 5.5: Market Centres in Southwest Iowa in the Summer of 1961.

Source: Berry (1967), p. 5. Reprinted by permission of Prentice-Hall Inc., Englewood Cliffs, NJ.

Figure 5.6: Southeast Iowa: Farmers' Preferences in 1964.

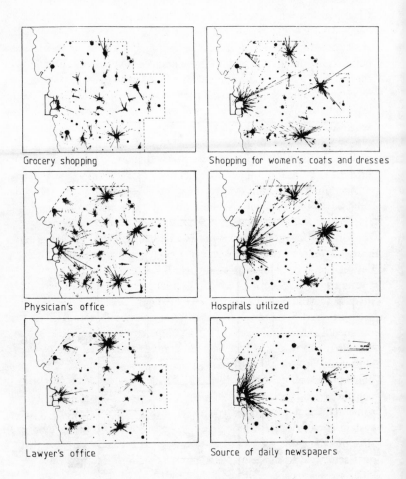

Grocery shopping

Shopping for women's coats and dresses

Physician's office

Hospitals utilized

Lawyer's office

Source of daily newspapers

Source: Berry (1967), pp. 11, 12. Reprinted by permission of Prentice-Hall Inc., Englewood Cliffs, NJ.

Despite variations imposed by local cultural and geographical circumstances, similar patterns and relationships have been identified in central place studies conducted in a number of areas throughout the world. In an early review, Barnum (1966) summarised findings which were common to analyses of functional hierarchies in Washington State (Berry and Garrison, 1958a), Iowa (Berry, 1967), Tasmania (Scott, 1964), New Zealand (King, 1962) and the Indian Punjab (Mayfield, 1967). These are summarised in Figure 5.7. The first is a close linear relationship between the population and the total number of retail establishments in centres. The second is an apparent curvilinear relationship between types of establishment and total establishments in each centre, consisting of several linear regimes which correspond to steps in the hierarchy of centres. The third is a curvilinear relationship between population and types of businesses. While by no means corresponding exactly to the patterns predicted by central place theory, these results broadly confirm the existence of a functional hierarchy of service centres.

According to the uniform spacing hypothesis, service centres at the same tier in the hierarchy should be spaced the same distance apart, and this distance will be greatest for the highest order centres. Empirical support for this proposition is provided by Brush (1953) who found the average distance between lowest order centres in Wisconsin to be 8.8 km, between middle order centres to be 15.0 km and between highest order towns to be 33.9 km. In a more detailed examination, King (1962) investigated the distributional characteristics of settlement in 20 sample areas in the US using nearest neighbour analysis. The technique, first outlined by Clark and Evans (1954), compares the actual with a theoretical distribution of points and the statistic ranges from 0.00 for a clustered pattern, through 1.00 for a random pattern, to 2.15 for a uniform (hexagonal) pattern. In 12 of King's sample areas, the nearest neighbour statistic was greater than 1.00 which indicates that there is some tendency towards geographical order within the central place system (Figure 5.8). Although this level of uniformity is far less than that predicted by the model, it does suggest that central place theory goes some way towards providing an explanation of the distribution of cities as service centres.

In an attempt to explain the discrepancies in central place theory as highlighted by these early empirical examinations, the model was elaborated and refined by relaxing some of its underlying assumptions and constraints. Most interest has centred around the behavioural restrictions implied in the concept of economic man, and these were removed so as to allow, more realistically, for imperfect information

Figure 5.7: Classical Central Place Theory: Population, Establishments and Central Functions Relationships.

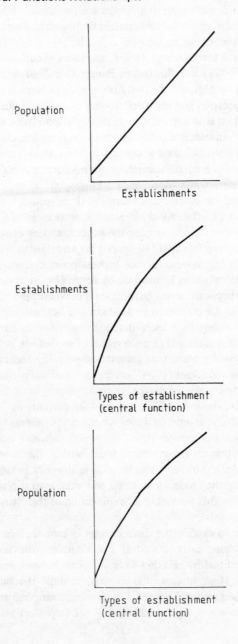

about prices and quality, and for the existence of consumer preferences not based wholly on distance. In Rushton's (1969) study of shopping patterns in Iowa, it was found that consumers do not always make their maximum purchases at the nearest available location. Distance and town size are important considerations, but distance to town decreases in significance as consumers shop for higher order goods and services. Moreover, there is evidence that consumers prefer to shop in the smaller of two equidistant cities, possibly because parking and traffic congestion are seen to be less of a barrier. What conditions spatial behaviour and hence determines spatial structure is consumers' perceptions of distances to places, and the range of goods, services and general facilities which they offer. Sufficient concensus exists among consumers for there to be a high level of order within the central place system, but the network, and the pattern of movements and purchases within it, is far less rigid and deterministic than classical central place theory suggests.

In a second reformulation, attempts have been made to allow for random variations, both in the physical landscape and of human behaviour, at the aggregate level. The former area of theoretical development is most closely associated with the work of Dacey (1966) who recast the locational basis of central place theory in a probablistic framework. Starting with a regular lattice of points of the type postulated by Christaller, Dacey formulated a statistical model which predicted the types of distortion which could arise under circumstances of varying relief. In testing the model against the distribution of towns in Iowa, he found a very good fit, but noted that this stochastic extension of the theory merely provided improved description, and added nothing by way of understanding of the underlying processes. For Curry (1964, 1967), the size, spacing and economic functions of service centres is determined by the interaction of the two groups of actors involved in the central place system, namely, the consumers and the retailers. The activities of the two groups involves certain random elements in both time (for example, the maintenance of stocks of goods by the retailer) and space (for example, the shopping at different centres by the consumer). In deciding upon the range and quantities of goods to stock, retailers are faced with the uncertain nature of demand which is associated with the unpredictability of consumers' behaviour. What reduces uncertainty is, however, the frequency of demand for goods and services over time, with the result that goods which are in constant demand, such as groceries, will be kept in ready supply, while holdings will be lower for items which are needed less commonly. Satisfying the total demand generated by a given population over a fixed period of time

Figure 5.8: Classical Central Place Theory: Nearest Neighbour Statistics for Twenty Study Areas in the United States.

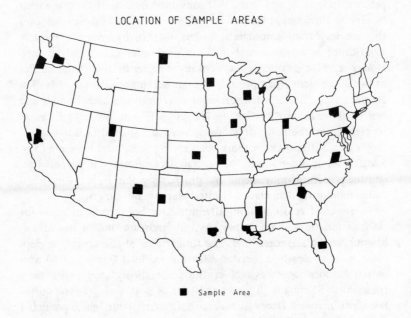

LOCATION OF SAMPLE AREAS

■ Sample Area

Near-neighbour statistics

Sample Area	Number of Towns	Density of Towns Per Square Mile	Mean Observed Distance (miles) rA	Expected Mean Distance in Random Distribution (miles) rE	Near-neighbour Statistic R	Nature of Pattern
California	96	0.0243	3.46	3.21	1.08	Random
Florida	64	0.0200	3.32	3.53	0.94	Random
Georgia	132	0.0350	3.52	2.67	1.32	Approaching uniform
Iowa	82	0.0307	3.86	2.85	1.35	Approaching uniform
Kansas	51	0.0166	5.16	3.88	1.33	Approaching uniform
Louisiana	140	0.0437	2.57	2.39	1.08	Random
Minnesota	55	0.0169	5.32	3.85	1.38	Approaching uniform
Mississippi	104	0.0280	3.84	2.99	1.28	Approaching uniform
Missouri	80	0.0219	4.67	3.38	1.38	Approaching uniform
New Mexico	23	0.0065	6.82	6.20	1.10	Random
North Dakota	28	0.0082	6.13	5.52	1.11	Random
Ohio	131	0.0512	2.80	2.21	1.27	Approaching uniform
Oregon	128	0.0317	2.86	2.81	1.02	Random
Pennsylvania	177	0.0466	2.28	2.32	1.22	Approaching uniform
Texas (NW)	38	0.0104	6.03	4.90	1.23	Approaching uniform
Texas (SE)	61	0.0182	4.29	3.70	1.16	Approaching uniform
Utah	20	0.0061	4.49	6.40	0.70	Aggregated
Virginia	122	0.0363	3.20	2.62	1.22	Approaching uniform
Washington	32	0.0073	4.14	5.85	0.71	Aggregated
Wisconsin	97	0.0299	3.58	2.89	1.24	Approaching uniform

Source: King (1962), p. 162.

leads to a network of centres varying in size and specialisation. The settlement structure can thus be seen as the product of randomly varying spatial and temporal behaviour in a random spatial economy. These contributions elevate central place theory to a highly symbolic and abstract form. They mean that it can be applied to areas with a complex physical geography, and where full knowledge of the market and rational spatial behaviour, both by individuals and overall, cannot be assumed.

A theory of urban location on very similar lines to that of Christaller was developed by Lösch in *The Economics of Location* (1954). Lösch's approach was altogether more complex because it considered not merely one organisational principle, but the combined effects of ten organisation principles operating at the same time. Lösch argued that every good sold and every service offered has a different range and threshold so that Christaller's $K = 3$, $K = 4$ and $K = 7$ networks were only limited cases of a progression of market area sequences that could also include K values of 9, 12, 13, 16, 19, 21 and 25 (Figure 5.9). Raising the number of organisational principles to ten increases the potential complexity of the settlement pattern enormously, but Lösch introduced order by arbitrarily centring all the meshes upon one point, which became the central metropolis. Further, these different networks were rotated so as to produce a maximum coincidence of points (the fewest number of centres). As a consequence of these procedures, some cities lie at the overlap of several markets, and so will perform many functions, while others are at the centre of only a small number of markets and so offer only a narrow range of goods and services. Lösch's economic landscape in fact consists of identical $60°$ sectors radiating from the central metropolis, each of which can be subdivided into two $30°$ sectors, one being 'city-poor' with few large cities, while the other has many large centres and is 'city-rich'. In contrast to Christaller's model, market areas are not nested and so there is a greater range of functional specialisation. Indeed, towns of the same size may perform very different economic functions, while centres of contrasting size may be economically similar. Christaller's condition that a centre at a given level will provide the goods and services of all lower order centres is not fulfilled with the single exception of the metropolis, so the 'hierarchy' is continuous and not stepped. The wide range of organisational principles involved suggests that the model is applicable to both service provision and to those manufacturing activities which are oriented towards, and are locationally determined by, their markets.

Central place theory can be criticised on a number of grounds. The

Figure 5.9: Lösch's Central Place System: Sectoral Arrangement of Networks of Market Areas.

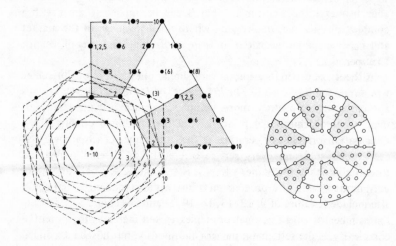

Source: Lösch (1954), p. 119.

most obvious reservations surround the restrictive nature of the underlying physical assumptions and behavioural constraints. Few places in the world exhibit the high level of topographic uniformity required by the model, and even on the North American Prairies, where flatness is the overriding feature of the landscape, the alignment of rivers and the lattice pattern of the roads which results from the division of the territory into rectangular sections, means that ease of movement in all directions is not possible. Similarly, omniscience and total rationality, the behavioural attributes of 'Economic Man', are highly unrealistic assumptions in a world in which most individuals act sub-optimally on the basis of imperfect information about distances and markets. One advantage of central place theory, however, is that it defines the ideal distributional patterns against which the actual locations of towns and cities can be compared, so as to identify the level of geographical efficiency. A second advantage is that it provides a basis for planning the size and spacing of centres in newly-developing areas. The major value of the theory is that, irrespective of its restrictive inputs, it does account for and explain many of the functional and distributional characteristics of service centres, as each of the studies listed in Table 5.2 demonstrate. That the fit is not perfect does not mean theory has no use and must be rejected. No simple theoretical statement can be expected to account for all aspects of a highly variable and complex retail servicing system. What central place theory emphasises is that systems of service centres are not totally unstructured and disordered. Regularity in the distribution of market centres, and in the functions they perform, exists because a sufficiently large number of retailers and consumers reach the same decisions about the supply and the consumption of goods and services.

Cities as Manufacturing Centres

Theories of the location of basic manufacturing activity trace their origins from the work of Weber (1909) and his concept of least cost location. Like Christaller, Weber made a series of simplifying assumptions about the physical landscape, and by ignoring considerations of labour costs and access to markets, focused primarily upon the costs of transport and raw materials as factors in location. He distinguished between weight losing (gross materials) and non-weight losing (pure materials) industries in order to determine minimum cost location. Paper making is an example of the former as the finished product is only about one-third the weight of the wood, rags and chemicals from which it is made, whereas brewing is weight gaining since bottled or barrelled beer is more weighty than the malt, hops, water and other ingredients that go into its manufacture. A further distinction was drawn between localised materials with restricted availability such as coal, iron ore, oil and chemicals, and ubiquitous materials such as water and air which are to be found everywhere. The former impose a major constraint on location; the effect of the latter is negligible.

The least cost formulation suggests that production activities locate so as to minimise overall transport costs, that is, the cost of assembling all the raw materials and dispatching the finished product to the market. Figure 5.10 presents a simple example in which there are two resource supply sites, R_1 and R_2, and a single market M. If an assumption is made of equal freight rates per unit of weight, the costs from each of these points can be represented by a series of equally spaced concentric and circular contours called isotims. Eash isotim describes the locus of points about each location where delivery or procurement costs are equal. Total transportation costs can be computed by summing the

Figure 5.10: Weber's Theory of Least Cost Location: Isotim and Isodapane Patterns According to Different Transport Cost Assumptions.

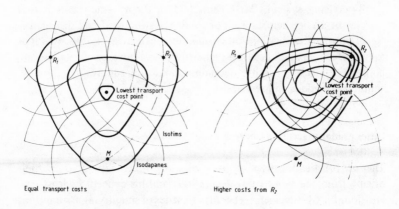

Equal transport costs

Higher costs from R_2

values of intersecting isotims. Lines connecting points with equal total transportation costs are termed isodapanes, and delimit the cost gradient for any commodity. In Figure 5.10(a), the least cost location is equidistant between the two resource points and the market. In reality, freight rates are likely to differ according to the type of transport used, and the nature of each slope will vary. For example, if movement costs from R_2 are twice those from R_1 and M this will be reflected in a different location of production (Figure 5.10(b)).

In historical terms it seems clear that the principles of least cost location were a major consideration in determining the distribution of manufacturing activities and hence of urban development. For example, the US iron and steel industry in the early nineteenth century was highly sensitive to assembly and distribution costs, on account of the weight losing characteristics of its two major materials – coal and iron. With early smelting techniques, more coal was required than iron ore so that coalfield locations like Pittsburgh and its surrounding towns were preferred. Later technologies drastically reduced the tonnages of coal needed and with the opening up of the Superior Uplands iron ore fields, centres of iron and steel making developed at least cost production points on the southern shores of Lakes Michigan, Erie and Ontario. Most recently, a set of north-east coastal locations for the US iron and steel industry were selected in response to the changing technology and costs of production combined with the increasing importation of ores from overseas (Warren, 1973).

The location of heavy manufacturing industry commonly generates further urban settlement associated with the growth of dependent, secondary economic activity. Opportunities are typically presented both for the supply of components, technical expertise and services, and for the use in ancillary industrial processes of by- and waste products, all of which increase the number of jobs and therefore the urban population in the area. Economies of agglomeration and linkage become more powerful as the size and diversification of industry increase so that industrial growth and associated urban development accelerate. Similarly, the expansion of urban markets attracts a wide range of retail and service activities so that further urban growth takes place. Examples of urban industrial complexes which have grown up around particular basic industries include the German Rühr (iron and steel), the Black Country of the English Midlands (metals) and Teesside (petrochemicals).

Although least cost location largely accounts for the distribution of basic heavy industry, and hence of the towns and cities which grew up with it, it is clear that for a great many manufacturing activities, raw material and transport costs are a comparatively minor locational consideration. Moreover, Weber's concern with only point markets, and his assumption that only one plant serves one market, seems highly unrealistic today. That access to the market is the overriding require-ment for much of industry was emphasised by Lösch (1954), who assumed in his model that production costs were the same for any location. His model therefore identified those places where, by gaining access to the widest markets, production would be most profitable. Both least cost and maximum market access principles were considered by Isard (1956) in an integrated theory of industrial location. He emphasised the importance of the substitution principle through which an optimal location can be achieved by increasing the outlay on a low cost factor of production to offset the burden of one that is more expensive. For example, given two equally advantageous sites for a factory, one may have cheaper land, the other cheaper labour. By locating on the cheaper land site, an entrepreneur would be substituting cheap land for cheap labour. For Greenhut (1956), this substitution is basically the same as the selection of a plant site from among alternative locations. In practice, the industrial location decision involves a compromise based upon a trade-off between all the raw material, labour, market access and transportation considerations that relate to the particular production process. That this compromise involves entre-preneurial decisions based, in the majority of cases, upon insufficient knowledge means that industrial location must be approached within a probablistic rather than a deterministic framework. Moreover, the domination of modern industrial production by a small number of very large multi-plant enterprises introduces a further consideration: what might appear to be the optimal location for an individual firm, might be sub-optimal within the context of the corporate empire. A final factor which is ignored in both Weberian and Löschian theory is the effect of government policy which, in attempting to direct the location of new industry, is in effect determining the pattern of future urban growth. These considerations emphasise the inherent complexity of the industrial location process. There can be no single theory of the location of industry, just as there can be no one theory of the location of towns and cities. The emergence of urban places as centres of manufacturing is explained by a set of raw material cost, market access, agglomeration and decision making principles which differ in their operation for each of the industrial activities which are present in the city.

The Location of Management Activities

The principles of location outlined in the preceding sections were primarily responsible for determining the basic distribution of towns and cities in most of the advanced countries of the Western world. The requirements of servicing and of manufacturing were all-important in the pre-industrial and early industrial economy, so that towns and cities located in those places which offered some combination of ease of access to markets and low cost or most profitable production. Recent changes in the structure of the economy, however, suggest that traditional central place and resource/market access considerations impose locational constraints on a declining number of urban economic activities. The most important trend, associated with the emergence of many Western nations as post-industrial economies, is the enormous increase in the number of jobs in geographically flexible management occupations. For example, a third of all economically active males in England and Wales are employed in professional, managerial and administrative occupations, and these were the most rapidly expanding, indeed the only growing, job sectors between 1961 and 1981. A second locational characteristic is the increasing control over the number and range of jobs in the city which is exercised by powerful firms and government institutions. To central place and industrial location theories must, therefore, be added a consideration of the principles which determine the location of corporate management.

The growth in management occupations is largely a result of the changes in the organisation of industry associated with the rise of the modern firm which were described in Chapter 3. For Hymer (1972), the small single product, single location family firm which characterised the early ninteenth century industrial economy was replaced first by the multi-department national corporation, and subsequently by the multi-divisional national and multi-national corporation. At each stage in this succession, an additional managerial tier was added to the firm so that both the total and the relative number of office-based jobs in manufacturing industry increased. Of particular importance was the emergence of multi-divisional forms in which corporations are de-centralised into several divisions, each concerned with one class of products, and each organised with its own head office. At a higher level, a general, group head or corporate office co-ordinates the divisions and plans for the enterprise as a whole. As well as a split between production and management, this development is associated with a separation of functions within management so that the divisional corporation is both organisationally and geographically most complex (Figure 5.11). A hierarchy of office-based management functions, spread across several countries and continents is indeed characteristic of the modern multi-divisional multi-national corporation.

The emergence of divisional forms is also associated with the increasing concentration of national production in relatively few enterprises. The multi-divisional structure was originated in the United States by General Motors and Du Pont between 1914 and 1920 and was first adopted in the UK by Imperial Chemical Industries (Reader, 1975), though there were many large companies organised on a departmental basis in 1939. Evidence of an increase in industrial concentration in the UK over the last sixty years is reviewed in Hannah (1976). The share of the largest 100 firms in manufacturing net output rose from 21 per cent in 1945 to 45 per cent in 1970. It is now nearer 60 per cent. Similarly, the proportion of assets in the quoted public sector of manufacturing and distribution held by the largest 100 firms increased from 44 per cent in 1953 to 62 per cent in 1963 (Utton, 1970). Over the same period, some seventeen of the manufacturing companies ranked in the largest 100 in 1953 were acquired by or merged with other manufacturing

Figure 5.11: Organisational and Locational Characteristics of Different Forms of Corporate Structure.

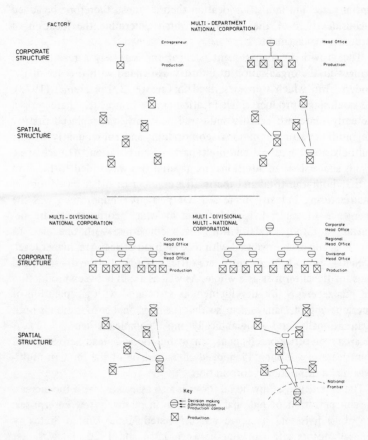

companies in the same group. Further evidence for concentration is found in the work of Samuels (1965) in which he argues that since the mid 1950s, larger firms have been growing faster than small ones thereby increasing the rate of concentration.

Focusing in more detail upon the organisation and control of modern manufacturing industry, Chandler and Redlich (1961) outline a simple scheme for classifying the management activities of multi-divisional national and international corporations. They distinguish three levels of business administration, three horizons and three levels of task, and three levels of decision-making and of policy formation. Level I, the lowest level, is designated production control and is concerned with managing the manufacturing operations of the firm. It involves those activities concerned with the receipt of raw materials, the dispatch of finished articles, and with organising production runs. Level II, which is the administrative level, is primarily responsible for co-ordinating the production controllers and is concerned with ordering, buying, marketing and with personnel management. The functions of level III, top management, are strategic and are to do with decision making. They involve goal determination and the shaping of the framework within which the lower management levels operate.

An important additional dimension is added to this typology by Thorngren (1970) who has pointed out that these organisational processes operate with respect to different time scales and generate different patterns of intra- and extra-firm communication (Table 5.3).

Table 5.3: Organisational Levels in the Multi-divisional Corporation.

Level of Management	Function	Activity	Time-scale	Pattern of Communication
—	Production	Manufacturing	Present	Intra-section
I	Production control	Programming	Short-term future	Intra-plant
II	Administration	Planning	Medium-term future	Intra-company (some external)
III	Decision making	Orientation	Long-term future	External and intra-company

The most important activities in a qualitative sense are those concerned with long-term scanning of socio-economic environments so as to identify future possibilities and alternatives for the business. They involve critical evaluations of future trends in raw material supply, technology and consumer demand, involving cost, political and legal considerations, and the making of strategic decisions as to which product lines to develop when. These 'orientation' functions are chiefly the responsibility of the highest level decision takers and are heavily dependent upon access to external sources of information about the likely future operating environment of the firm. They involve contacts with the directors of firms which are major suppliers and purchasers, and with individuals knowledgeable about the state of stock markets, the prosperity of national economies, the direction of government policies and developments in science and research. At a lower level, the administrative activities of the corporation are concerned with 'planning' the development of the firm in the directions that have been identified through higher level orientation processes. They are responsible for ensuring that the firm or division has the right number and mix of workers available and that contracts for raw material purchase and product sales have been placed in order to ensure the success of future production runs. Many of the contacts arising out of these activities are internal to the firm, involving the receipt of directives from level III and the issuing of instructions to production managers, though the medium-term perspectives of level II activities requires the maintenance of a basic network of external relations with suppliers, purchasers and consultants so as to synchronise forward planning schedules. Finally, level I activities are almost exclusively concerned with 'programming' the present activities of the firm and so largely involve internal channels of communication. Matters to be dealt with such as the receipt of raw materials and the dispatch of finished products occur on a regular basis and tend to involve production managers in consultation with shop floor workers.

Although this occupational classification is developed in the context of the industrial firm, it has a wider relevance for public sector employment. National government, like big business, is organised on a divisional basis with the executive branch and the individual ministries having similar types of responsibilities and functions as group head and divisional head offices respectively. Decision making functions in UK government are performed by ministers, and senior civil servants at the Secretary level. Individual government 'departments' are in fact analogous to corporate product divisions, having specific responsibility for major areas of policy such as education, health and welfare, environment, industry, employment, agriculture and trade and development. Parallels with production and production control can be found in the regional and local offices of national government 'departments' where staff and officers are concerned with routine dealings with the public.

Given the different functions performed at these levels and the different patterns of contact which they generate, it may be expected that the different economic functions will assume different spatial distributions (Table 5.4). Industrial location theory suggests that programming activities which control day-to-day production will spread

Table 5.4: Corporate Functions: Locational Preferences

Level	Function	Locational Requirements	Geographical Pattern
I	Production; production control	Least cost/maximum market access sites	Resource/market oriented location
II	Administration	Access to controlled information	Clustered: medium-sized cities
III	Decision making	Access to random information	Concentrated: central business districts of major cities

themselves over space, along with production units which themselves locate according to the pull of markets and raw materials. They will locate alongside manufacturing plants in locations which offer the best opportunities for minimum cost or most profitable production. Administration, because of its need for white collar workers, some external communication and professional services, will tend to concentrate in medium-sized cities. Since their demands are similar, and because it is in their common interests, corporations from other industrial sectors and branches of commerce will place their co-ordinating offices in similar sized cities so that administration will be far more geographically clustered than production and production control. Medium-sized cities will therefore offer a range of jobs in both production, production control and administration. Orientation functions must be even more concentrated since they must be located as close as possible to the capital market, government and research institutions. Their preference is for the central areas of the very largest cities since it is these locations which offer the best access to uncontrolled and random information sources which are essential for determining the long-term strategy of the firm. Distinctive patterns of land use and interaction may be expected in the downtown areas of these major cities as a consequence of the concentration of corporate power and control. Prestige buildings housing the group head offices of national and multinational corporations, branches of government, finance houses and employers and trades federations will dominate the skyline. The complex will be held together by a dense network of business contacts through which decision makers will interact on a face-to-face basis.

Research into the locational distributions of administration and management has been comprehensively reviewed by Daniels (1975, 1979) and Goddard (1975). At the national level, occupational patterns have been mapped for functional urban areas in Britain for 1971 by Drewett *et al.* (1976). Drewett's analysis broadly confirms the previous arguments by outlining the basic contrasts in the occupational structure of British cities (Figure 5.12). As expected, professional and managerial (level III) occupations are heavily concentrated in London, whereas skilled, semi- and unskilled manual occupations, which are the nearest census equivalents to production jobs, are most numerous, and therefore dominate in percentage terms, in the urban-industrial SMLAs of the North, Midlands, Scotland and South Wales. Despite the distinctiveness of planning activities in corporate management, however, their spatial separation from decision making functions appears to be incomplete as intermediate non-manual or administrative occupations are almost as concentrated as professional and managerial jobs. An important regional component characterises the distribution of occupations in urban Britain. London and the South East have a monopoly of top management jobs; conversely, British provincial cities are almost wholly deficient in administrative and decision making occupations.

The concentration of industrial control in London is further em-phasised by Goddard and Smith (1978). Of the top 1,000 companies in the United Kingdom, 525 had head offices in the Greater London Planning Region in 1977. Moreover, a location in the capital is particu-larly important to the largest firms as 304 of the top 500 had their head office there as opposed to 221 of the firms ranked 501-1,000. Important evidence of an increase of concentration was also revealed in this analysis. Greater London gained 19, and the Southeast Planning Region (with the exception of Greater London) gained 17 head offices between 1972 and 1977, whereas losses were recorded by the Northwest (15), the East Midlands (10) and Yorkshire and Humberside (9). Only minor changes in the number of head offices present were recorded for the remaining regions.

One major locational attraction of the central area of the very large city is the wide range of business information and intelligence generated by the very large number of firms and agencies which are present in the area. Patterns of inter-firm linkage have been monitored and analysed by Goddard (1973) using data on telephone calls reported by decision makers and administrators working in central London. In classifying these office activities, Goddard identified six major groupings concerned, respectively, with commodity trading, publishing and business services, civil engineering, fuel and oil, official agencies and banking and finance, and found that these accounted for very nearly two-thirds of the

Figure 5.12: The Occupational Characteristics of Urban Britain in 1971. The maps show the percentage of all economically active males in different social groupings, by standard metropolitan labour area (SMLA).

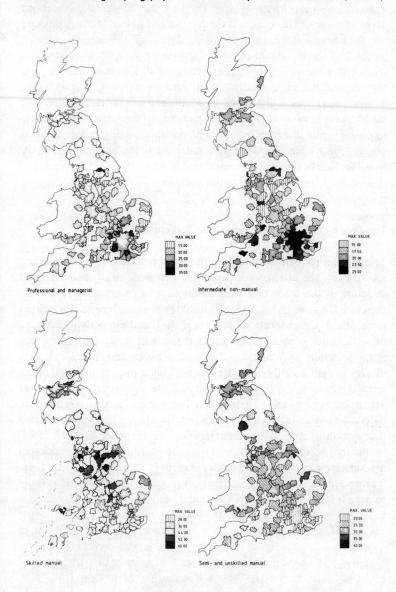

Source: Drewett (1976), p. 35. Reproduced by permission of the authors.

variation in the overall pattern of calling. Two particular features of the contact network were noted. The first is that all firms within a specialist group are not linked together by reciprocated information flows: in addition to within-cluster linkages, there are also important connections between groups. The second is the hierarchical nature of many linkages which, combined with the interdependency of the various clusters, means that the individual contacts form part of a highly integrated and structured system. The strength and complexity of connections suggests that the advantages of location in the area are substantial and are self-reinforcing. Firms which open head offices in Central London can clearly 'plug into' a rich and diverse network of contacts which is likely to enhance the effectiveness of their decision making processes. Conversely, firms which leave the area and move to the provinces are likely to suffer 'communications damage' and to find that the effectiveness of their strategic orientation functions declines.

The Urban Landscape: Evolution and Structure

Taken together, the locational arguments outlined in the preceding sections amount to a general theoretical statement on the evolution and structure of the urban pattern in advanced urban economies. Moreover, they provide an explanation for some of the historical material presented in chapter 4. Urban location theory suggests that the underlying network of settlements in an urban-industrial region will be arranged on central place principles in which the distances between and functional complexity of centres will reflect, historically, the transport and rural-urban servicing needs of the area (Figure 5.13). Different resource endowments across the region, and in particular the distribution of coal, iron ore, metals and chemical deposits, account for the superimposition upon this pattern of specialised manufacturing centres or complexes located, in theory, according to Weberian least cost and Löschian maximum market access principles. In practice, except in remote areas where there was no existing settlement, manufacturing activity tended historically to attach itself to the nearest appropriate service centre thereby distorting the central place pattern. Production activities are the dominant feature of factory firms and the very small numbers employed in 'management' performed, in effect, production control functions. Moreover, the single product, single location nature of the early capitalist economy was characterised by a distribution of independent, 'free-standing' urban manufacturing centres.

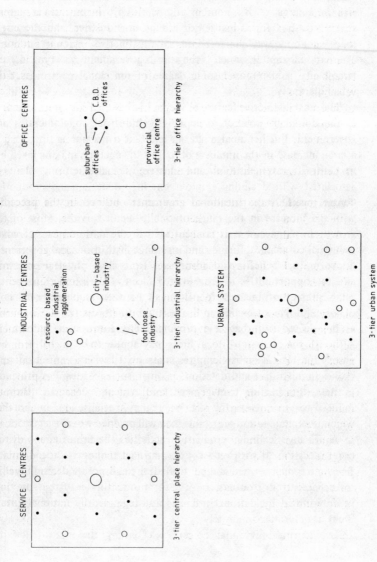

Figure 5.13: Functional Specialisms within a Hypothetical Urban System.

A different industrial/urban pattern is associated with the high capitalist economy which emerged in the late nineteenth century. The formation, through mergers and take-overs of firms with similar product specialisms, of national organisations organised on a departmental basis, increased the numbers of administrators in industry, whereas levels of employment in production and production control functions stablised. An important feature of corporate organisation at the time was the emergence of a number of powerful provincial cities as administrative centres for industries or groups of industries, and the urban space economy was characterised by a distinctive 'regional' structure. The national capital, which is the seat of government, was typically the largest city, and imposed some degree of functional control over the urban hierarchy.

The most distinctive feature of the multi-divisional corporate economy is the dominant position occupied by industrial conglomerates and government. Further amalgamations and take-overs among firms lead to a large increase in the numbers of jobs at the decision making level, but in contrast, rationalisation and closure of manufacturing plants is associated with declining employment in production and production control tasks. An expansion of government bureaucracy is associated with the increase in the range of public sector services. Most of this growth in management is concentrated in the national capital where industrial companies, finance and insurance institutions and government derive mutual benefits and advantages from close physical proximity and the opportunities it provides for face-to-face interaction and the establishment of business 'confidence'. As a consequence, centres of office employment emerge in the central business district (CBD) as well as in selected suburban and provincial city locations (Figure 5.13). Although the identity of local firms may appear to be unchanged, their absorption into industrial empires is associated with a centralisation of both administrative and decision making responsibilities. For provincial capitals, this loss of management employment means that they are reduced to a manufacturing role. Where they specialise in manufacturing which is expanding and profitable they will prosper; where they specialise in nationally declining production activities the urban economy will enter recession. If with respect to a national economy, the majority of firms are foreign owned and controlled, then corporate decision making is exercised from overseas, and the form of the urban system, previously determined by national interests and regulated by national government, is determined remotely.

In the multi-divisional corporate economy the mix of activities present determines the occupational structure, and through wages and salaries, the rewards for labour, the socio-economic environment of cities. Only the largest cities have a complete occupational profile by virtue of their attractiveness to production, production control, administration and decision making. The residents of these cities have access to the best jobs which typically involve the allocation of men and money at the highest level, and planning growth and development, tasks which carry the highest rates of remuneration. Conversely, small centres offer the most restricted range of job opportunities as they are primarily attractive to production and prodution control functions. Income differentials within such centres are more confined, and social mobility tends to be equated with geographical mobility as personal advancement is contingent upon migration to large labour markets which offer a wider range of better jobs. Middle ranking centres have a more balanced occupational profile though they lack many of the top jobs. As a consequence of intermediate size, however, they offer an attractive balance between urban amenity and access to the countryside, so avoiding the worst excesses of pollution and congestion experienced in larger centres.

Although the evidence and the explanation are much disputed, it appears that the urban landscape, as it evolves, is characterised by important changes in the size distribution of cities. The ranking of cities was first discussed in the literature in the inter-war period when it was observed by Zipf (1949), among others, that there was in many countries a regular gradation of cities according to size. In fact the second city was invariably, according to Zipf, half the size of the first, the third city a third the size of the first and so on, so that the size of any centre could be predicted simply by its rank and the population of the largest place. Thus:

$$P_n = \frac{P_1}{r_n}$$

where P_n is the population of the nth city, P_1 is the population of the largest city, and r_n is the rank of the nth city. For example, if the largest city has 10 million population, then the tenth ranking city will have one million population. A fundamentally different, primate distribution was, however, observed in other countries by Jefferson (1939). In this pattern there are one or two very large centres and the remaining settlements are small in size.

For Berry (1961) these different size distributions are sequentially related and reflect the changing economic and political complexity of an economy. From an investigation of city size distributions in 38 countries, he noted that rank size regularity appeared to be typical of larger countries such as China, Brazil, the USA, India and South Africa and of countries such as Italy, Belgium, Switzerland and West Germany which have long urban traditions. Smaller countries, and those with shorter histories of urban development, such as Denmark, Sweden, the Dominican Republic and Sri Lanka tended to be dominated by a single large centre and so to have a primate pattern. A third category of nations was found to exhibit an intermediate size distribution on account of the absence, or small number of cities, in particular size classes. For example, Australia was included in this group because small cities were missing while a large proportion of the country's population lived in large centres. England and Wales are also assigned to this group because of the absence of middle size cities. Similarly, Canada was included because of the lack of a dominant centre at the top.

These observations pointed to the relationship between the type of size distribution and the number, direction and strength of forces acting upon each unit that was first noted by Simon (1955). For Berry (1961), 'primacy is the simplest city size distribution affected by but few simple strong forces . . . At the other extreme, rank size distributions are found where, because of complexity of economic and political life and/or age of the system of cities, many forces affect the urban pattern in many ways' (p. 582). In simple economies, one or two forces act strongly upon the urban system so that a single city, commonly the national capital, grows to dominance (Figure 5.14). Examples of primate cities thus include the capitals of small countries historically engaged in the primary production of relatively few commodities as is the case of Vienna (Austria), Stockholm (Sweden), Amsterdam (Netherlands) and Copenhagen (Denmark); ex 'colonial' cities in underdeveloped countries such as Colombo (Sri Lanka), Bangkok (Thailand) and Montevideo (Uruguay); and former centres of empire such as Madrid (Spain) and Lisbon (Portugal). As economic complexity increases, so does the range of urban specialisms, so that there are many more stimuli to urban growth, and with the increasing size of intermediate and smaller centres, a rank size pattern emerges. Such complexity is clearly greatest in the largest most advanced nations although countries such as Australia and Canada which are poorly integrated internally, or those which are actually physically divided such as New Zealand and Pakistan (in 1961), have yet to advance their urban systems beyond an intermediate size distribution. A basic problem with this analysis and explanation is clearly

Figure 5.14: Berry's Developmental Model for City-size Distributions.

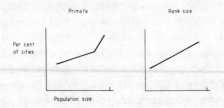

Types of city size distribution

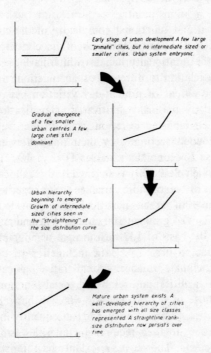

Stages of development

Early stage of urban development. A few large "primate" cities, but no intermediate sized or smaller cities. Urban system embryonic.

Gradual emergence of a few smaller urban centres. A few large cities still dominant

Urban hierarchy beginning to emerge. Growth of intermediate sized cities seen in the "straightening" of the size distribution curve

Mature urban system exists. A well-developed hierarchy of cities has emerged with all size classes represented. A straightline rank-size distribution now persists over time

Source: Berry (1961), p. 583. Reproduced by permission of the University of Chicago Press.

that political boundaries, especially in Europe, from where many of Berry's examples were drawn, have shifted continually over time so introducing difficulties in simply relating urban size patterns to national historical and developmental factors. Moreover, the size of the country itself is in many cases rather poorly depicted by its population or geographical area. What the study means, however, is that in the course of development, the city size distribution in a country shows a gradual progression from a primate, through an intermediate, to a rank size pattern. Such changes are not a simple reflection of the level of economic development that has been achieved, they relate instead to the range, strength and direction of the forces which affect the growth of cities. Rather than become more complicated as these forces proliferate and intensify, the level of internal order increases.

The Contemporary Urban System

The locational and functional complexity of the contemporary urban system reflects the varied histories of its constituent cities. It is a product of the locational principles which have operated with varying intensities in relation to different urban functions at different times in the past. The extent of present-day variation between cities is amply demonstrated by the many multivariate classifications of urban places that have been reported in recent years. In multivariate classification, cities are grouped according to their similarities as measured over a number of socio-economic variables (Berry, 1972). No single index decides to which class a city is assigned, rather, allocation is determined by the effects of two or more variables acting together.

An example of urban classification is provided by the work of Armen (1972) who grouped 100 cities in England and Wales (excluding London) on the basis of 132 measures of demographic, locational and activity characteristics. The data included statistics on population; employment; climate; finance; administrative, educational, recreational and sporting facilities; and retail and social functions. The procedures used were those of cluster analysis which is basically a technique for combining cities, first into pairs, and subsequently into groups and sub-groups according to their reciprocal similarities as expressed by correlation coefficients. The most important characteristic of the groups is that while not wholly discrete, constituent members are more similar to each other than they are to members of other groups. The final step in the analysis was the ranking of values which identified the most prominent characteristics of each class of cities. This ranking identified the dominant activity system in each combination and was used to demonstrate that urban specialisation meant a systematic variation of social, economic and environmental characteristics which endowed an area with a sense of place. This variation of characteristics led to a nomenclature based upon the distinctive activities of each group.

Armen's taxonomic scheme consisted of six major classes of city, each of which comprised places which had similar environments, socio-economic profiles and opportunities for work and play. Although places may be included in more than one class, the patterns themselves are very different in geographical terms (Figure 5.15). For example, a relatively even distribution of administrative/market centres throughout the country contrasts markedly with the concentration of industrial centres in the North and Midlands, the latter pattern reflecting the historical importance of raw material and market access factors. The importance of south coast cities for holiday and retirement activities is emphasised as is the contribution of the New Towns to the 'rapidly expanding towns' pattern. The varied geographical character of the urban system is underlined. Different locational principles combine to produce specialised urban centres which in turn form the elements in a complex urban assemblage.

What integrates these cities into an urban system is the linkages between them. It is the pattern of functional interdependencies which both makes possible, and is a reflection of, the differences in urban socio-economic character. Contacts between cities take a variety of forms, and include movements of people, goods and ideas. Daily commuting and shopping movements are examples of the first group, and by linking places of residence with points of employment and service provision, they both make possible, and are necessitated by, locational specialisation. Distance imposes major constraints upon these movements and so labour markets and retail trade areas are typically compact and highly centralised. Such movements are among the most well understood of all types of flow as the earlier discussion of central place theory and related work has shown.

By connecting resource points to manufacturing centres, and manufacturing centres to markets, commodity flows introduce strong ties of economic interdependency into the urban system. Lengths of haul of raw materials and finished products vary widely according to the weight gaining/weight losing characteristics of the commodity. For Ullman (1962) the principles of complementarity, intervening opportunity and transferability determine the structure of commodity flows. For centres

Figure 5.15: A Classification of Cities in England and Wales, 1966.

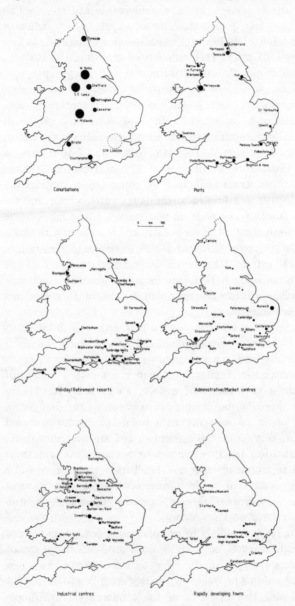

Source: Armen (1972), pp. 170-3.

to interact, there must be a supply or surplus of resources in one place, and a corresponding demand or deficiency in another. Complementarity will, however, generate flows between pairs of places only if there is no intervening centre in a position to serve as an alternative source of demand, and if shipment costs, or transferability arrangements, are acceptable. To these principles must be added a consideration of location since it is distance which determines the mode, and hence in many cases, the feasibility of shipment. For example, over short hauls, road and rail will be preferred, while on long routes, sea and air transport normally prove most cost effective. With modern telecommunications, information and ideas can be exchanged regardless of distance and directional considerations so they provide effective links between major cities and the smallest and most remote centres. These arguments suggest that, although apparently chaotic and confused, urban linkages are likely to be as highly structured and as explicable as the familiar variations in urban size and function.

Despite their equivalent importance, urban linkages in general have been studied in a far less comprehensive manner than urban socio-economic characteristics. Restrictions of data mean that except in local area studies, most workers have been limited to a single measure of connection so providing only partial insights into functional structure. There are thus individual investigations of airline traffic (Taaffe, 1956, 1962), telephone calls (Clark, 1973) and financial flows (Borchert, 1972), but few studies have investigated the effects of several types of exchange acting together.

A recent exception to this generalisation is Davies and Thompson's (1980) analysis of the flows of 15 commodities among 17 major centres in the Canadian Prairies. The methodology used was the principal components analysis approach as developed by Berry (1966) which identifies both major groups of commodities, and patterns of directed flow between places. The analysis showed that two-thirds of the total variation in traffic could be explained in terms of three underlying linkage patterns. Component one, which accounted for 29 per cent of the total explanation was termed a tertiary and heavy industries dimension as it grouped together general and household goods and commodities associated with manufacturing activities. Figure 5.16 shows the structure of this pattern in which the data values denote the relative importance of each link. For example, the Winnipeg-Calgary dyad has a score of 4.4 and the Calgary-Winnipeg dyad a value of 2.2 which indicates that the Winnipeg-Calgary flow is twice as important as the reverse flow for the commodity groups associated with this axis. The pattern is indeed dominated by

Figure 5.16: Patterns of Connectivity Among Urban Centres of the Canadian Prairie Provinces as Revealed by Factor Analysis of Commodity Flows.

(a) Component I: Tertiary and Heavy Industry (29.4%)

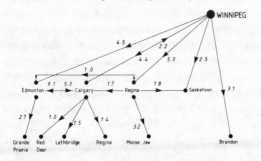

Commodities	Loadings
General Freight	0.88
Metal Products	0.81
Machinery	0.79
Construction	0.77
Household	0.70
Others	0.62
Foodstuffs	0.49*
Petroleum	0.40

(b) Component II: Food Products and Light Manufacturing (22.3%)

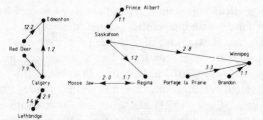

Commodities	Loadings
Trailers	0.91
Perishable Food	0.87
Foodstuffs	0.68
Seed and Feed	0.60
Livestock	0.54*
Machinery	0.50*

(c) Component III: Raw and Semi-processed Materials (15.1%)

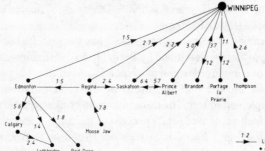

Commodities	Loadings
Livestock	0.72
Minerals	0.69
Forest	0.58
Chemicals	0.57
Petroleum	0.33*

$\xrightarrow{1.2}$ Link and Component Scores ≥ 1.0

* Secondary Loading

Source: Davies and Thompson (1980), p. 306.

Winnipeg, with Edmonton, Calgary and Regina forming a second tier of places which in turn link to the subservient centres of Grande Prairie, Red Deer, Lethbridge, Regina and Moose Jaw. Twenty-two per cent of the variance is accounted for by a food product and light manufacturing component. The pattern of scores on this factor defines two separate sub-graphs, one based in Alberta, the other in Saskatchewan and Manitoba. Four commodity classes have their highest loading on component 3, which accounts for 15 per cent of the variance. Although superficially similar to component 1, this pattern is altogether more complex. One notable feature is the focusing of flows from several centres upon Winnipeg, the other is the partial independence of the Alberta cities. Although the overall pattern of connectivity is highly involved, this study highlights the underlying structural order within the Prairie urban system. These patterns are both a product of, and make possible, the urban hierarchy in the region.

Beyond the Analytic Approach: Modelling Urban Systems

The preceding sections of this chapter have depended heavily upon an analytical approach to the study of urban systems. This has involved reducing the system to its constituent cities and linkages and then investigating each of these disaggregated components in turn. One focus of attention has been upon the principles which determine the distribution of towns and cities and the ways in which they combine to form a theoretical statement on urban location. The other has been upon the controls and constraints which limit urban interaction. Empirical studies which have examined cities and their linkages have enabled these explanations to be refined. Urban systems appear to have an ordered structure, much of which can be accounted for by locational and linkage theory.

Although it has identified basic relationships, it is apparent that the analytical approach is not without very serious drawbacks. The most severe limitation is the sheer impracticality of isolating each individual component of a complex urban system for specific investigation, and of reconstituting the insights gained into an overall perspective. It is for this reason that many urban geographers have turned to modelling urban systems at the aggregate scale. Rather than analyse individual cities and linkages it is the system as a whole that is the central focus of attention.

In systems modelling, a system is defined as a series of elements that can take alternative states. For example, in an urban system, each city consists of a series of parcels of land which are assigned to different uses. These are determined by the landowner with reference to his needs and to the characteristics of the site. Where the land can only be used for residential or industrial purposes, the range of alternative states is small, but the greater the number of options open to the individual landowner, the greater is the potential complexity of the overall urban pattern. A simple analogy for this very obvious statement, as suggested by Gould (1972), is provided by the tossing of an unbiased coin. In a single toss, there are two possible alternative states, heads and tails. With two tosses there are three (two heads, a head and a tail and two tails), while with three tosses there are four alternative states (three heads, a head and two tails, two heads and one tail, and three tails). Note that the order is unimportant (a head and a tail is the same as a tail and a head); it is the number of different configurations that matters.

Although a number of alternative states are possible, an important characteristic of urban systems, indeed of any system, is that one set of states, or configuration, is likely to dominate. Moreover, the more complex the system, in other words the greater the range of possible alternative states, the more one particular configuration will dominate all the others. This characteristic is explored in Table 5.5 in the context of a very simple land use system in which each plot of land in the city can only be used for residential or industrial purposes. If the city is divided into only two plots, there are three possible land use configurations, although the pattern of one residential and one industrial exerts some dominance. Where there are four plots, the total number of possible combinations rises to 16, but the emergence of one small set of land use patterns becomes clear. Increasing the number of plots to six, raises the total number of combinations to 64. The system is, however, characterised by considerable bunching around the dominant configuration, as the percentages show. Although any one of the 64 configurations could occur, a small set of configurations has a much greater likelihood of occurrence than any other. This propensity of one particular configuration to overwhelm all the others is known as the tendency towards the most likely state. It becomes more powerful, the more complex the system. Despite the potentially chaotic structure which might be expected to follow from a large mass of elements and linkages, it suggests that any urban system is likely to display a particular and distinctive form.

It is this property that forms the basis of urban systems modelling. The aim is to identify the most likely state of the urban assemblage given information not about individual elements and links, but about

Table 5.5: The Emergence of a Dominant Configuration in a Simple Land Use System.

	Combination of States	Overall Configuration of System	Frequency of Occurrence of Configuration	Occurrence as Percentage of Dominant Occurrence	Characteristics of Distribution
Two plots					
	RR	2R	1	0.50	
	RI,IR	1R,1I	2	1.00	Some dominance
	II	2I	1	0.50	
Four plots					
	RRRR	4R	1	0.17	
	RRRI,RIRR, IRRR,RRIR	3R,1I	4	0.67	
	RRII,RIRR,RIIR RIRI,IIRR,IRIR	2R,2I	6	1.00	Some bunching around dominant configuration
	RIII,IIIR IIRI,IRII	1R,3I	4	0.67	
	IIII	4I	1	0.17	
Six plots					
		6R	1	0.05	
		5R,1I	6	0.30	
	Total of 64 different combinations	4R,2I	15	0.75	
		3R,3I	20	1.00	Considerable bunching around dominant configuration
		2R,4I	15	0.75	
		1R,5I	6	0.30	
		6I	1	0.05	

Notes: R = residential land use; I = industrial land use.

Source: after Gould (1972).

the macroscopic characteristics of the system as a whole. Gould (1972), reviewing Wilson's (1970) seminal work in this field, illustrates the problem by reference to a simple system consisting of two residential or origin areas, and three workplace or destination areas, in which only the overall distribution of seven homes and seven jobs is known. In this example, there are 36 possible trip distributions, and the problem is to find that trip configuration which overwhelms all the others and therefore represents the most likely state of the system. The solution is one in which both row and column totals equal 7: no workers can be assigned to a zone in which all the housing is occupied or work in an area in which all the jobs are taken. Additional constraints can be introduced in the form of total travel costs so as to ensure that the commuting pattern is not financially unrealistic. The aim is, therefore, to find the most likely configuration of the system given fixed residential and job distributions, and overall cost constraints.

Solutions to this problem rely heavily upon entropy maximising techniques as developed by Wilson (1970). Entropy is a measure of the degree of order in a system, and maximising techniques identify the most ordered, or most likely state. As the equations are large, and there are so many unknown values, the mathematics are formidable, and involve the use of Lagrangian multipliers and combinational methods which are unfamiliar to most geographers. They are explained in simple terms by Gould (1972). Large, high speed computers enable the various values to be estimated and the distribution of trips to be compared with the pattern known to exist in the area. In this way the model developed by Wilson *et al.* (1969), predicted very accurately (0.95) the distribution of journey to work movements by modes, among towns in Northwest England.

The primary importance of the modelling approach is methodological, and lies in the fact that the structure of an urban system can be specified without the need for a detailed analysis and understanding of the properties and characteristics of individual elements and links. This does not mean that the distinguishing features of cities and their traffic patterns are unimportant and can be ignored; rather, that at the aggregate systems level, where a very large number of centres and flows is being considered, the variations associated with individual elements and links are self-cancelling and are subsumed by the macroscopic properties of the system. Modelling provides a powerful means of analysing urban systems which parallels and complements studies of cities and their traffic flows at the individual level.

A second advantage is the applied or practical value of urban systems modelling. Once an urban system has been summarised in mathematical form and has been calibrated against contemporary data, the model can be run for different dates in the future so as to predict, for example, the patterns of inter-urban movement that are most likely to be associated with the growth of cities in the system. Models for predicting future patterns of movement typically consist of four sub-models concerned with trip generation, trip distribution, modal split and assignment. 'Generation' is used as a generic term for 'production' and 'attraction' and so this part of the model predicts the total number of trips originating in each zone and the total number of trips finishing in each zone. Estimates are derived from empirical data on household characteristics and economic activities, the assumption being that different social and occupational categories give rise to different volumes and directions of movement (Wootton and Pick, 1967). The trip distribution model predicts how the trips leaving a zone will be distributed among attraction zones, that is, among all other zones. The modal split model then allocates proportions of each bundle of movements to the major transport modes such as car, bus, train and bicycle, based upon assumptions about transport preference relating to distance of movement and the socio-economic character of the person making the trip. The final stage involves the assignment of flows to the road and rail network so as to show both the volume and type of traffic which is expected on each route in the future. On this basis, transport engineers and economists can make decisions about investment and construction. Examples of the use of this methodology for transport planning purposes include the SELNEC (South East Lancashire-North East Cheshire) Transportation Study (Wilson *et al.*, 1969, Wagon and Wilson, 1971) and the Manchester Area Rapid Transit Study (1967).

Although applications in the field of transport planning were the first to be developed, systems modelling techniques have a far wider relevance in urban geography. A third feature of the approach is indeed the general utility of entropy maximising approaches for analysing distributional and allocational processes and patterns at the aggregate level. The methodology has been used, for example, to estimate demographic changes (Rees and Wilson, 1977), to predict shopping patterns (Openshaw, 1975) and to specify urban land uses (Batty, 1976). Given this wide relevance, this review provides merely the briefest and most elementary outline of the methodological and technical bases of modelling approaches. Urban systems modelling is a sophisticated area of inquiry which constitutes a major field of study in advanced urban geography.

Conclusions

At first glance a modern urban system, consisting of a set of cities varying in size and economic function and interconnected by a range of commodity, population and information flows, presents a confusing picture to the analyst. Some places are major centres of political and corporate power, others are dynamic and prosperous metropolises, while others are stagnating manufacturing cities in declining industrial regions. The system is also likely to include a varied array of satellite, dormitory, religious, transport, recreational, resort and retirement centres. Despite this functional complexity, urban geographers, by focusing upon particular aspects of the urban economy, have developed a general theoretical understanding of the distribution of towns and cities as service, manufacturing and management centres. Similarly, principles and theories of interaction account for many aspects of urban inter-linkage. When integrated and combined in an historical context these identify and account for a high level of locational and organisational order in the urban system.

Urban systems are geographically structured in a number of complex ways. One form of order is locational and exists because the different elements in each city's economy are arranged according to basic principles of distribution. Availability of raw materials, access to markets and levels of demand mean that the incidence of economic specialisation varies considerably so that cities exhibit different occupational profiles and related socio-economic characteristics. The effects of space are also reflected in the patterns of linkages between centres which appear to correspond to the principles of distance decay, complementarity, intervening opportunity and transferability. A second source of order is hierarchical and refers to the size distribution of cities. In simple agricultural service economies, a stepped hierarchy of centres is normally apparent, but with the introduction of an industrial and manufacturing economy in which the influences which determine the size of cities are many and varied, and act randomly, a continuous, rank-size pattern tends to emerge. The third regularity is developmental and relates to the geographical components of change which in most urban systems take place in the form of waves of diffusion spreading outwards from central poles of innovation. Geographical order is both a cause and a consequence of the prevailing modes of technology in the urban economy. This means that fundamentally different arrangements of the urban system have arisen over the last two centuries in association with the early industrial, the late industrial, the corporate and the post-industrial economies.

6 THE INTERNAL STRUCTURE OF THE CITY

One of the most characteristic features of modern cities is their high level of internal differentiation. Sets of zones, communities or neighbourhoods are normally distinguishable in terms of physical appearance, population composition and related social characteristics and problems, and these are repeated from one city to another. The existence of similar social and residential patterns suggests that urban structure is determined by a number of general principles of land use and location. It points to the operation of powerful underlying social and economic forces which encourage similar, if not identical, uses for adjacent parcels of land across the city. The identification and explanation of internal patterns and processes is a major topic of inquiry in urban geography.

A number of approaches have been followed by urban geographers in their attempts to understand urban spatial structure (Table 6.1). The first is essentially ecological and seeks to account for urban patterning in terms of the struggle for location and space in the city. It places particular emphasis upon compeition over territory between social groups and the ways in which this leads to the emergence of 'natural' areas in each centre. The trade-off approach is grounded in neo-classical economics and explains intra-urban land use patterns in terms of the outcome of competitive bidding for land. Social area analysis and factorial ecology are extensions of the basic ecological approach which interpret intra-urban structures in terms of general theories of social and economic change. The conflict/management approach has roots in political science and the analysis of power and conflict in the city. It investigates the institutional structure of land use and development with reference to the role of urban managers and the nature of the constraints imposed upon individuals and groups within the city. The final approach is explicitly Marxist, and explains the existence of social and spatial divisions in the city in terms of the capitalist organisation of society. It outlines the way in which the landowning class impose and manipulate rents so as to ensure the most profitable (for them) geographical arrangement of land uses in the city. The internal structure of the city is examined in this chapter in the context of these six broad perspectives.

141

Table 6.1: The Internal Structure of the City: Alternative Analytical Approaches.

Approach	Theoretical Background	Areas of Inquiry	Major Contributors
Ecological	Human ecology	Struggle for space among human groups	Park (1916); McKenzie (1925)
Trade-off	Neo-classical economics	Utility maximisation; rent bidding	Thunen (1826)
Social-area analysis	Urbanisation	Social consequences of societal development	Shevky and Bell (1955)
Factorial ecology	Factor analysis	Social and spatial patterns in the city	Berry (1971)
Conflict/management	Weberian sociology	Power groupings; 'gatekeepers'	Cox (1976); Pahl (1975)
Marxist	Historical materialism	Urban land use theory; housing allocation mechanisms	Harvey (1973)

Source: adapted from Bassett and Short (1980), p. 2.

The Ecological Approach

The ecological approach to the study of urban communities is most closely associated with the writings, between 1916 and 1940, of the Chicago School of Urban Sociology. Members of the school viewed the city as an object of detached sociological analysis, and sought to explain the complexities of the urban community and to discover patterns of regularity in its apparent confusion. Very similar objectives with regard to plant and animal communities were being pursued at that time by biologists in the context of the new science of ecology which emphasised the interdependence of specimens and species, and the relation of each kind and individual to its environment. 'Human ecology is therefore the study of the spatial and temporal relations of human beings as affected by the selective, distributive, and accomodative forces of the environment' (McKenzie, 1925, p. 64). For Park (1936), human society is organised into two levels — the 'natural' or biotic and the 'novel' or cultural. It is at the former level that ecological processes, similar to those which determine the structure of plant and animal communities, operate. These impersonal forces which relate to man as a species, rather than to man as a repository of beliefs and values, work their way through the social system to create patterns of social differentiation in the city.

Familiarity with their own city led the Chicago ecologists to the recognition of orderly patterns of social structure, and of change, in the character of urban neighbourhoods. In the most abstract representation, Burgess (1925) reduced Chicago to a number of concentric rings, arranged about the central business district, each containing a number of 'natural areas' (Figure 6.1). The extension of central commercial uses into surrounding residential districts, or the movement of immigrant groups into the inner city, it was argued, triggered off predictable patterns of displacement and movement of population through the rest of the urban area and provided the basic mechanism of urban change. Explicit use in the analysis of structure and change was made of ecological terminology. The area of 'dominance' in the city is its commercial centre which is equivalent to the dominant species in a plant community and which sets the conditions for the existence of subordinate species; in this case, the city's social groups. Populations or activities 'invade' territories displacing previous occupants until they 'succeed' to the control of that space. The resulting 'segregation' of the population, although clearly related to land values is, therefore, equivalent to the symbiotic balance maintained by plant and animal species in an ecological community and is, therefore, natural in derivation.

Figure 6.1: Burgess's Concentric Zone Model of Urban Social and Spatial Structure.

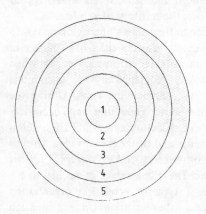

BURGESS' CONCENTRIC ZONES:

1   Central business district

2   Transition zone

3   Independent workingmens homes

4   Better residences

5   Commuter zone

These arguments, and in particular their ecological bases, are best illustrated by reference to the work of McKenzie (1925). He maintained that as a community grows there is not merely a multiplication of people and buildings, but differentiation and segregation as well. Residences and institutions spread out in centrifugal fashion from the central part of the city, while business concentrates more and more around the point of highest land value. Each cyclic increase of population and businesses compounds the struggle for the best locations in the city, and as competition intensifies, the economically weakest are forced out to the less accessible and less valuable areas. Over time, the central point emerges as a clearly defined business area dominated by banks, department stores and hotels. Industries and factories usually comprise independent formations within the city, grouping around railway tracks and routes of water traffic. Residential areas become established, segregated into types dependent upon the economic and social composition of the population.

For McKenzie, the structural growth of the city took place in a successional sequence not unlike the successional stages in the development of plant formations. Certain specialised forms of uses do not appear in the city until a particular stage of development has been achieved, just as the beech or pine forest is preceded by successional dominance of other plant species. And just as in plant communities, successions are the products of invasions; so also in the city, social patterns are the result of the formations, segregations and associations which appear when one set of uses encroaches upon the territory of another.

Invasions are classified according to their stages of development into initial, secondary and climax. In rapidly growing cities like Chicago in the early twentieth century, the initial stage of invasion occurs when one social group attempts to expand geographically and encounters resistance by those residents in the surrounding areas. During the development stage, keen competition between newcomers and occupants stimulates a process of displacement, selection and assimilation determined by the character of the invader and of the area invaded. Groups forced out attempt to move into other areas, and so trigger-off further invasion and succession sequences. The climax stage is reached in the invasion process once a dominant type of ecological organisation emerges which is able to withstand further intrusion. Under a situation of repeated invasion and succession, a set of well-defined communities develops in the city. In the language of plant ecology, these were termed 'natural areas' by the Chicago sociologists.

As the products of complex ecological processes of competition, invasion-succession and assimiliation-segregation, natural areas define the basic social divisions in the city. They represent the habitats of groups of people who, on the basis of existence in a common territory, develop distinctive traditions, customs and conventions. They are 'natural' in that they are not the result of design, but rather a manifestation of the ecological tendencies inherent in the urban structure. For Park (1925), natural areas were important both as methodological tools which enabled field workers to generalise from their own research to other areas of the same kind, and in a more practical sense, as the basis for community organisation within urban society. The natural area concept was developed furthest in the highly perceptive studies of Zorbaugh (1929) and Wirth (1928). *The Gold Coast and the Slum,* a detailed descriptive analysis of North Side Chicago, enabled Zorbaugh not only to identify the major sub-areas in the region, such as Hobohemia and Little Sicily, but also to advance explanations of their formation that were somewhat more profound than the simplistic ecological arguments that, in general, characterised the work of the Chicago school. The ecological approach was valuable in that it focused upon competitive interactions and interrelationships, but by equating man with members of the plant and animal kingdoms it ignored the economic, institutional and political forces that also contribute to the formation of social patterns in cities.

Economic Approaches: The Trade-off Model

Economic theories of urban structure trace their origins to the agricultural land use model first proposed by Thunen in 1826. The basic premiss in Thunen's approach was that decisions as to which crop to grow where are determined by profitability, which is a function of sale price at, minus costs of production and shipment to, the farm gate. The most profitable crops are grown around the farm, and the least profitable in the most distant fields, so the farmhouse is encircled by rings of land use of declining intensity and value. Similar arguments underlay the trade-off model of urban structure as originally proposed by Hurd (1903) and Haig (1926), and as restated and elaborated more recently by Kain (1962), Alonso (1964), Muth (1969) and Mills (1972). The basic hypothesis is that property owners attempt to minimise location costs by trading off rents against travelling costs. Differences of bid-rent function between major groups of land users give rise to concentric circles of social and economic activity in the city (Figure 6.2).

Figure 6.2: Rent-distance Relationships in the Bid-rent Theory of Urban Structure.

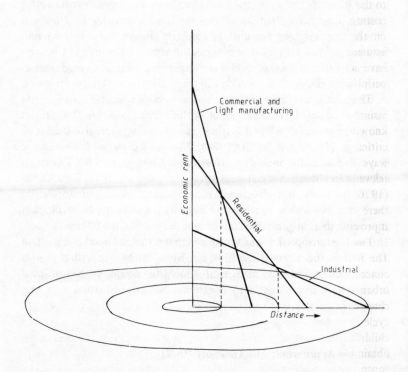

The trade-off model relates to a highly idealised set of locational and behavioural conditions. Assumptions are made of a monocentric city situated in a featureless plain with all the jobs located in the city centre. Within the city, transport costs increase directly with distance away from the centre but rent, which is dependent upon accessibility alone, varies inversely with distance. Users of land in the city make locational decisions within the limits of their total budgets. For those in high quality retail or commercial activities, accessibility to the population is all important so that gradient showing the amount of rent which they are prepared to pay falls off steeply with distance from the centre. Low income groups minimise transport costs by locating near to the centre but as this is also an area of expensive housing, savings are made by

living at high densities. Their bid-rent gradient has, therefore, an intermediate slope. High income groups have the option either to live close
to the centre so reducing travelling costs but having to face high housing
costs, or to take advantage of cheaper, better, more spacious housing
on the periphery and to bear the high journey-to-work costs. It is
assumed that high income households prefer low density living so they
have a shallow bid-rent gradient which expresses a preference for
peripheral locations.

This elementary trade-off model is necessarily simplistic. It involves
assumptions of a monocentric city, perfect competition, complete
knowledge of the market and freedom of choice, which are open to
criticism. In recent years, the trade-off model has been reformulated in
ways designed to overcome these shortcomings and to increase its
relevance to the complex modern city as Evans (1973) and Romanos
(1976) have shown. Indeed, as Richardson (1977) suggests, since 1970
there has been such a resurgence of interest in trade-off approaches that
it provides the basis of a 'new urban economics'.

Two features of the simple trade-off model are worthy of attention.
The first is that it provides, retrospectively, a rationale for Burgess's
concentric zones, and so fuses ecological and economic explanations of
urban social structure. The second is that an explanation of urban
change is provided by linking bid-rent functions to stages in the family
cycle. For example, if a larger house is required when a couple have
children, they are expected to move further out from the centre to
obtain more space at lower cost. Conversely, when household size
contracts as children leave home, there is a corresponding movement
back towards the centre to minimise expenditure on travel and accessibility. These links between expenditure, social circumstance and urban
structure are explored more fully in social area analysis explanations of
geographical differentiation in the city.

Social Area Analysis

The study of urban social and spatial structure was advanced significantly
in the 1940s by the development of an approach which came to be
known as social area analysis. The technique was initially outlined by
Shevky and Williams (1949) as a method for classifying census tracts,
but was re-presented by Shevky and Bell in 1955 as a model of social
change which amounted to a general theory of urban differentiation
(Table 6.2). The model is based upon three broad conceptual notions

Table 6.2: Social Area Analysis: Steps in Construct Formation and Index Construction.

Postulates Concerning Industrial Society (Aspects of Increasing Scale) (1)	Statistics of Trends (2)	Changes in the Structure of a Given Social System (3)	Constructs (4)	Sample Statistics (Related to the Constructs) (5)	Derived Measures (from Col. 5) (6)
Change in the range and intensity of relations	Changing distribution of skills:	Changes in the arrangement of occupations based on function	Social rank (economic status)	Years of schooling Employment status Class of worker Major occupation group Value of home Rent by dwelling unit Plumbing and repair Persons per room Heating and refrigeration	Occupation Schooling Index Rent I
	Lessening importance of manual productive operations; growing importance of clerical, supervisory, management operations				
Differentiation of function	Changing structure of productive activity	Changes in the ways of living; movement of women into urban occupations; spread of alternative family patterns	Urbanisation (family status)	Age and sex Owner or tenant House structure Persons in household	Fertility Women at work Single-family Index dwelling units II
Complexity of organisation	Lessening importance of primary production; growing importance of relations centred in cities, lessening importance of the household as economic unit				
	Changing composition population:	Redistribution in space; changes in the proportion of supporting and dependent population; isolation and segregation of groups	Segregation (ethnic status)	Race and nativity Country of birth Citizenship	Racial and national groups in Index in relative III isolation
	Increasing movement; alterations in age and sex distribution; increasing diversity				

Source: Murdie (1971), p. 284.

concerning the changing character of modern society: changes in the range and intensity of relations, differentiation of function and increasing complexities of organisation. These postulates are grounded respectively in the work of Wilson and Wilson (1945) on increasing scale or inter-dependence, on Clark's observations (1940) concerning the division of labour in society, and on Wirth's (1938) examination of the relationship between population concentration and social form (Parkes, 1972).

Shevky and Bell argued that the increasing scale of modern urban industrial society was associated with basic changes in economic and social relationships. Changes in the range and intensity of social activities gave rise to a new economic order requiring different types of skills and jobs, so that occupation and education became important indices of social differentiation. At the same time, differentiation of function increased the range of social and economic options available in the city and made possible a number of alternative patterns of activity to the traditional family-centred life style. The most important reflection of an increasingly complex social order was seen as the changing mobility and composition of the urban population, and the cultural isolation of many ethnic groups. Shevky and Bell did not claim that the social areas they devised were either the natural areas of the human ecologists or communities in the sense of intense interaction. Rather, the social area generally 'contains persons having the same level of living, the same way of life, and the same ethnic background, and we hypothesize that persons living in a particular type of social area would systematically differ with respect to characteristic attitudes and behaviours from persons living in another type of social area' (p. 20). From these arguments are derived the three basic constructs of increasing scale: social rank (or economic status), urbanisation (or family status) and segregation (or ethnic status). A range of variables indexing these constructs was provided from which six were selected for detailed analysis.

Social area analysis proceeds to identify sub-areas in the city by mapping a composite standardised index ranging from 0 to 100 for each construct. For example, the social rank index consists of an education and an occupational measure, and individual census tracts are assigned to a low, middle or high position on a scale based on the values they show on these two variables. Tracts are next evaluated on the basis of an urbanisation index as measured by fertility, proportion of women in the labour force and the percentage of housing in family units. Finally, a segregation index is calculated by determining the distribution of the population in designated birthplace groups. This methodology has been applied in several studies of the social structure of cities throughout the world (Table 6.3).

Table 6.3: Social Area Analysis: Selected Studies.

Country	City	Study
USA	Los Angeles	Shevky and Williams (1949)
	San Francisco	Shevky and Bell (1955)
	Akron, Dayton, Indianapolis, Syracuse	Anderson and Egeland (1961)
UK	Newcastle-under-Lyme	Herbert (1967)
Canada	Winnipeg	Herbert (1972)
Italy	Rome	McElrath (1962)
Australia	Newcastle	Parkes (1972)
	Melbourne	Jones (1969)

Figure 6.3 shows an application of the social area technique to Coventry. Differences between the UK and US censuses mean that a

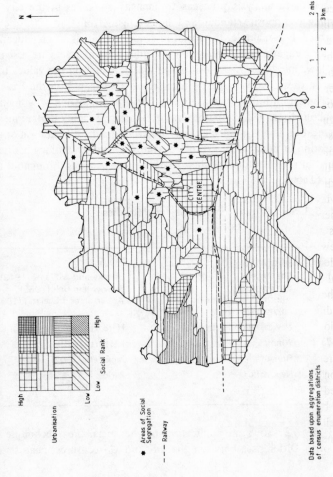

Figure 6.3: Social Area Analysis as Applied to Coventry, 1971.

CITY CENTRE

Urbanisation

High

Low

High   Low
Social Rank

*   Areas of Social
    Segregation

--- Railway

Data based upon aggregations
of census enumeration districts

N

0   1   2 mls
0   1   2   3 km

slight adaptation of Shevky and Bell's proposed methodology was necessary. This was achieved by defining social rank in terms of occupation ratio (total number of semi-skilled and unskilled workers per 1,000 employed persons) and education ratio (number of economically active persons with education below ONC or 'A' level per 1,000 persons economically active). Urbanisation was indexed by a fertility ratio (children aged 0-4 in relation to females 15-45) and a women at work ratio (women at work in relation to women aged 15-45). Segregation was based on the percentage of New Commonwealth, Old Commonwealth, Europe, Ireland and 'other' foreign born per area. Segregation is highly concentrated to the north of the central area, as the distribution of this variable picks out those parts of the city in which immigrant groups are concentrated. Otherwise, the overall social area pattern is somewhat confused: most zones occupy a midway position along the urbanisation and social rank scales and there are few cases of over- or under-scoring on these measures. One reason for this indifference may be the particular socio-economic character of Coventry, which as a relatively prosperous, upper working class city, lacks the extremes of social rank and urbanisation. The other is possibly that social area analysis, as a technique designed to identify social areas in large North American cities of thirty years ago, is only of limited diagnostic value in the contemporary middle-sized UK city. Evidence to support this latter point is restricted and limited. Social area analysis has undoubtedly advanced the understanding of the social and spatial structure of the American city as the individual studies listed in Table 6.3 demonstrate, but other than Herbert's 1967 study of Newcastle-under-Lyme, applications to UK cities have been rare. Newcastle-under-Lyme provides, moreover, a somewhat questionable basis from which to generalise. It is a much smaller city (population 75,000), and like Coventry has a distinctive social structure, lacking, in this case, a coloured immigrant population.

The most important criticism of social area analysis is that Shevky and Bell did not answer the fundamental question of how the major social divisions in the city determine spatial structure. Hawley and Duncan (1957) argued in a detailed critique that there is no way of relating social differentiation, which is a product of increasing scale, with a real differentiation at the census tract level, and that the model is therefore characterised by inherent structural weaknesses. Although not originally stated, links between social theory and geographical patterns can be identified as Bell (1965) and Murdie (1969) have shown. These suggest that the internal structure of cities will be characterised by rings of family status, wedges of socio-economic status and clusters of ethnicity (Figure 6.4).

Figure 6.4: Social Area Constructs: Theoretical Patterns.

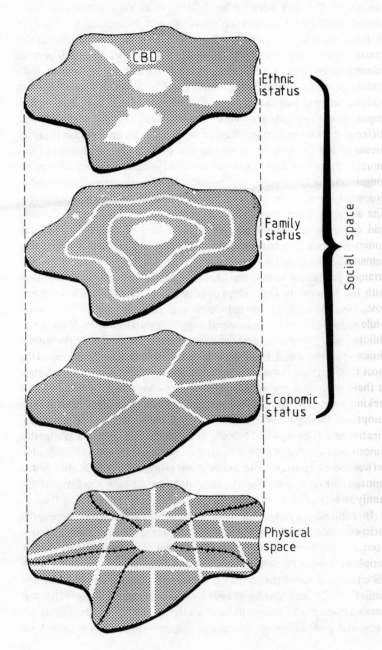

For Bell, a different set of locational preferences is associated with the three alternative life styles, termed familism, consumerism and careerism which he identified in the city. Despite a continuing decline in popularity, a life style oriented towards marriage and child rearing remains the choice of many. Familism falls into a number of stages, relating broadly to age, which include pre-marriage, marriage pre-children, child rearing, child launching, marriage post-children and widowhood, each of which generate different locational and residential requirements. For example, the housing needs of the unmarried will differ considerably from those of couples with young children who will presumably require more space, and rank access to parks and schools more highly. Similarly, those whose children have left home will no longer need large houses, and with increasing age, may prefer smaller properties requiring less upkeep and located in more accessible positions. The ability to pay for property also varies according to age and those in mid career, while requiring more housing space for their families, are commonly best placed financially to acquire it. Variations of housing demand and levels of disposable income suggest that familists will arrange themselves in concentric rings around the city in accordance with the bid-rent surface. The young and the old will prefer more low cost, small unit and high accessibility locations close to the centre, while the suburbs will be characterised by nuclear families of parents and children. This basic pattern is likely to be reinforced by the locational choices of consumerists and careerists. Bell argued that those who place most emphasis upon work would prefer central and inner city locations, as these offer best access to jobs in the downtown area. Similarly, those seeking personal expression through materialism and conspicuous consumption would be most assisted by being close to central shops, theatres and entertainment centres. In this way, life style choice generates concentric patterns in the city. A progressive graduation is projected between the heterogeneous inner area, occupied by the old, young unmarried adults, consumerists and careerists, and the socially uniform family suburb.

In contrast to concentric rings, a sectoral pattern is projected for socio-economic status in the city. The arguments here largely derive from the work of Hoyt (1939) who, while recognising basic centre-periphery contrasts, identified spoke-like wedges of land use in many US cities. These sections were explained in terms of a concentration of similar uses around radial access routes. For example, railways and canals attract heavy industry, whereas freeways and parkways generate linear residential developments. Aversion to industry, a preference for high land and areas close to parks and open space are resolved by the development of sectors of socio-economic status.

Cultural differences finally suggest that ethnic groups will be both a socially and spatially discrete component in the city. Language, religion and custom provide a powerful basis for social segregation and the need to be close to meeting houses, religious centres, specialist shops and places of entertainment limits the geographical spread of ethnic communities. Income and occupation are additional constraints and restrict the immigrant at least initially to those areas in the city which offer both jobs and cheap housing. These conditions define an expectation of clusters of ethnicity in and around the central area of the city. Rather than being homogeneous and uniform, these arguments suggest that social space in the city will consist of a combination of concentric, sectoral and clustered patterns.

Empirical support for these hypotheses was provided by Anderson and Egeland's (1961) study of the social geography of Akron, Dayton, Indianapolis and Syracuse. They used analysis of variance techniques to test whether the distribution of construct score values in these cities was patterned or could have arisen by chance. The survey design was such that the existence of sectoral patterns for social rank scores and zonal patterns for urbanisation scores were tested. Anderson and Egeland's tests of statistical significance showed that, as predicted, social rank had a wedge-shaped expression, while urbanisation was concentrically distributed. Similar findings were reported by Herbert (1972) for Winnepeg. Although the geographical characteristics of the constructs have received far less attention than their socio-economic complexion, these studies do provide considerable support for Bell's arguments. The highly selective nature of the input data, however, is a serious shortcoming. There remains the possibility that mapping of composite scores based upon other variables might reveal geographical patterns that could not be accounted for by social area theory.

Factorial Ecology Approaches

Criticisms of the operational limitations of social area analysis, though initially severe, were largely bypassed in the early 1960s by the increasing availability of high speed computers which made possible a more sophisticated examination of census data on the city. In place of the six simple measures of urban social structure as originally proposed by Shevky and Bell, a wide range of socio-economic indicators could be considered. Moreover, their interrelationships could be explored by multivariate correlation and classification techniques. By far the greatest use has been made of factor analysis, and particularly of principal components analysis, as a means of identifying the social and spatial structure of cities. The term factorial ecology is generally used to describe these approaches.

Factorial ecology is a purely technical procedure. It is essentially taxonomic and provides a means of identifying the latent patterns which exist within a multivariate data set. Unlike social area analysis, there is no theoretical framework, and so no direct inferences can be drawn as to the nature of the processes which give rise to the social and spatial patterns which are revealed. The approach is illustrated by a study of the social and spatial structure of Royal Leamington Spa, Warwickshire (population 50,000) based on the 1966 census (Figure 6.5). As the basis for analysis, 41 indices, which were believed to be indicative of the major patterns of socio-economic variation, were abstracted for each of the 20 enumeration districts (census tracts) in the city so as to form a raw data matrix (Table 6.4). The first stage in principal components analysis involved the examination of basic associations among the data through intercorrelation. Each variable was compared with every other variable so as to produce a 41 x 41 inter-correlation matrix. The technique then identified least squares linear combinations of variables, known as components, within this matrix subject to the twin constraints of maximum variance reduction and othogonality. The mathematics are explained in simple terms by Rummell (1967) and in more detail by Horst (1965). Forty-one components, which are essentially new vectors of variables, were abstracted, but as a consequence of the criteria employed, the leading components accounted for, or explained, the majority of the variance in the data. In this case, five components accounted for 75 per cent of the input variation.

This initial components structure can be simplified by redistributing the explained variance among the five abstracted patterns through a process of rotation, to produce a more specific solution. Table 6.5 shows the five components which result after varimax rotation. Each component or vector consists of 41 loadings with values between +1 and -1 which show the ways in which the input variables combine in the form of basic patterns of association within the data. The empirical meaning of these dimensions is revealed by an examination of the leading loadings on each component.

Figure 6.5: Factorial Ecology of Royal Leamington Spa, Warwickshire: Major Stages in the Analysis.

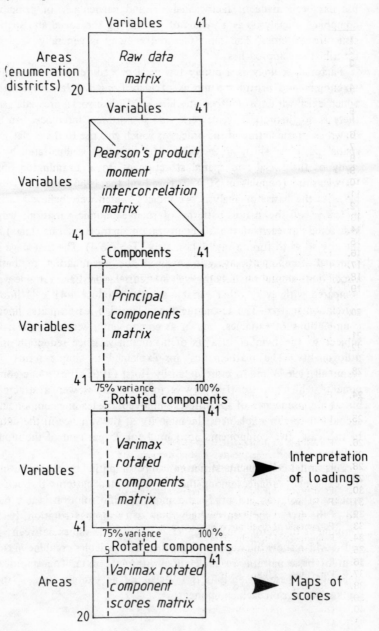

Table 6.4: Factorial Ecology of Royal Leamington Spa, Warwickshire: List of Variables.

---

1. Total population.
2. Economically active females as percentage of all females.
3. Economically active males in government as percentage of all economically active males.
4. Economically active males in social classes I and II as percentage of all economically active males.
5. Economically active males in social classes IV and V as percentage of all economically active males.
6. Percentage of cars garaged within curtilage.
7. Percentage of persons living at a density of over 1½ per room.
8. Percentage of households lacking a fixed bath.
9. Percentage of households sharing a dwelling.
10. Percentage of dwellings which are multi-dwellings, purpose-built.
11. Percentage of dwellings which are multi-dwellings, converted.
12. Percentage of the total population commonwealth born.
13. Percentage of the total population foreign born and born at sea.
14. Percentage of the total population born in Ireland.
15. Percentage of the total population who are single.
16. Percentage of the total population aged 0-4 years.
17. Percentage of the total population aged 15-24 years.
18. Percentage of the total population aged 65 years and over.
19. Percentage of the total population moving within local authority area during the last five years.
20. Percentage of the total population moving into local authority area during the last five years.
21. Economically active males as percentage of all males.
22. Economically active males in construction as percentage of all economically active males.
23. Economically active males in services as percentage of all economically active males.
24. Economically active males in socio-economic groups 1, 2, 3, 4, 13 as percentage of all economically active males.
25. Economically active males in socio-economic groups 7, 10, 11, 15 as percentage of all economically active males.
26. Percentage of households with no car.
27. Percentage of households with two or more cars.
28. Percentage of persons travelling to work by bus.
29. Percentage of persons travelling to work by car.
30. Percentage of persons travelling to work on foot.
31. Percentage of dwellings which are owner-occupied.
32. Percentage of dwellings which are rented from local authority.
33. Percentage of dwellings which are privately rented.
34. Percentage of households with no family.
35. Percentage of households with six or more persons.
36. Percentage of persons living at a density less than 1½ per room.
37. Percentage of households with one to three rooms.
38. Percentage of households with seven or more rooms.
39. Percentage of households with exclusive use of all amenities.
40. Overall sex ratio (male/female).
41. Persons per room.

Table 6.5: Factorial Ecology of Royal Leamington Spa, Warwickshire: Varimax Component Loadings Matrix.

Variables		Components				
		1	2	3	4	5
1.	Total population	0.24	0.18	0.15	−0.25	−0.
2.	% economically active females	0.64	0.30	0.20	−0.29	0.
3.	% economically active males in government	−0.43	−0.33	−0.17	−0.47	0.
4.	% economically active males in social classes I and II	−0.76	0.26	−0.51	−0.09	−0.
5.	% economically active males in social classes IV and V	0.76	0.06	0.22	0.33	−0.
6.	% cars garaged in curtilage	−0.84	0.00	0.05	0.14	0.
7.	% population density of over 1½ per room	0.12	0.04	0.59	−0.28	−0.
8.	% households lacking fixed bath	0.49	0.35	−0.25	0.47	−0.
9.	% households sharing a dwelling	0.01	0.80	−0.23	−0.08	−0
10.	% multi-dwelling, purpose-built	−0.14	−0.02	0.11	0.06	0
11.	% multi-dwelling, converted	0.04	0.85	−0.14	−0.25	0
12.	% commonwealth born	0.42	−0.11	−0.21	0.68	−0
13.	% foreign born and born at sea	−0.29	0.61	0.04	0.44	−0
14.	% born in Ireland	0.32	0.77	−0.05	0.02	0
15.	% single	0.20	−0.27	0.80	−0.20	0
16.	% aged 0-4 years	−0.00	0.04	0.03	0.86	0
17.	% aged 15-24 years	0.02	0.06	0.07	−0.50	−0
18.	% aged over 65 years	−0.25	0.40	−0.42	0.25	−0
19.	% moving within local authority 1961-6	0.11	0.28	0.09	0.10	0
20.	% moving into local authority 1961-6	−0.32	0.55	−0.64	0.09	−0
21.	% economically active males	0.28	0.34	−0.51	−0.46	−0
22.	% economically active males in construction	0.51	−0.51	0.37	0.41	0
23.	% economically active males in services	−0.44	0.63	−0.34	−0.27	−0
24.	% economically active males in socio-econ. groups 1, 2, 3, 4, 13	−0.72	0.25	−0.52	−0.18	−0
25.	% economically active males in socio-econ. groups 7, 10, 11, 15	0.77	−0.04	0.23	0.38	−0
26.	% households with no car	0.78	0.16	0.37	−0.16	−0
27.	% households with 2 + cars	−0.72	−0.10	−0.29	−0.11	−0
28.	% persons travelling to work by bus	−0.23	−0.11	0.70	0.20	0
29.	% persons travelling to work by car	−0.87	−0.02	−0.31	−0.02	0
30.	% persons travelling to work on foot	0.83	−0.00	−0.19	−0.07	−0
31.	% dwellings owner-occupied	−0.57	−0.22	−0.42	0.48	−0
32.	% dwellings local authority rented	0.25	−0.40	0.66	−0.39	−0
33.	% dwellings privately rented	0.31	0.76	−0.43	−0.02	−0
34.	% households with no family	0.11	0.84	−0.41	−0.08	−0
35.	% households 6 + persons	0.34	−0.36	0.69	−0.02	
36.	% population density below 1½ per room	0.07	0.22	−0.80	0.05	−
37.	% households with 1-3 rooms	0.18	0.91	−0.03	−0.01	
38.	% households 7 + rooms	−0.29	0.14	−0.51	0.27	
39.	% households with exclusive use all amenities	−0.66	−0.16	0.20	0.06	
40.	% overall sex ratio	0.03	0.36	−0.05	−0.58	−
41.	Persons per room	−0.37	0.10	−0.83	0.03	
	Per cent variance explained	22.10	17.30	17.10	10.30	7
	Cumulative per cent explained	22.10	39.40	56.50	66.80	7

Component 1 was interpreted as socio-economic status dimension. The high positive loadings group together a number of measures of low class and socio-economic positions, while the negative loadings identify affluence and status in terms of car and property ownership (Table 6.6). Variables to do with birthplace, mobility and family and household structure which are indicative of a transient bed-sitter land population gave component 2 a distinctive character. Component 3 is the most complex and difficult to interpret, but the polarisation of property ownership, residence size and associated social characteristics points to the existence of a housing type and tenure dimension in the data. Components 4 and 5 account for smaller amounts of the variance, and were interpreted as ethnic status and migration measures respectively. Despite the enormous demographic, occupational, residential and mobility variations in the city, the analysis points to the existence of five basic dimensions within the social structure of Royal Leamington Spa. These are highly similar, in both size and complexion, to those identified by Evans (1973) in his study of Swansea, Newport and Cardiff.

The geographical characteristics of these dimensions are expressed by the rotated components scores which are calculated for each of the areal divisions (census enumeration districts) of the city (Table 6.7).

Table 6.6: Factorial Ecology of Royal Leamington Spa, Warwickshire: Leading Component Loadings.

Component 1: 'Socio-economic status'		(22.1 per cent explanation)
Number	Variable Name	Component Loading
30	Percentage of persons travelling to work on foot	+0.83
26	Percentage of households with no car	+0.78
25	Percentage of economically active males in socio-economic groups 7, 10, 11, 15	+0.77
5	Percentage of economically active males in social classes IV and V	+0.76
2	Percentage of economically active females	+0.64
31	Percentage of dwellings which are owner occupied	−0.57
39	Percentage of households with exclusive use of all amenities	−0.66
27	Percentage of households with two or more cars	−0.72
24	Percentage of economically active males in socio-economic groups 1, 2, 3, 4, 13	−0.72
4	Percentage of economically active males in social classes I and II	−0.76
6	Percentage of cars garaged within curtilage	−0.84
29	Percentage of persons travelling to work by car	−0.87

Component 2: 'Bed-sitter land'	(17.3 per cent explanation)	
Number	Variable Name	Component Loading
37	Percentage of households with one to three rooms	+0.91
11	Percentage of dwellings which are multi-dwellings, converted	+0.85
34	Percentage of households with no family	+0.84
9	Percentage of households sharing a dwelling	+0.80
14	Percentage of the total population born in Ireland	+0.77
33	Percentage of dwellings which are privately rented	+0.76
23	Economically active males in services	+0.63
13	Percentage of total population foreign born and born at sea	+0.61
20	Percentage of total population moving into local authority area during the last five years	+0.55

Component 3: 'Housing type and tenure'	(17.1 per cent explanation)	
Number	Variable Name	Component Loading
15	Percentage of the total population who are single	+0.80
28	Percentage of persons travelling to work by bus	+0.70
35	Percentage of households with six or more persons	+0.69
32	Percentage of dwellings which are rented from local authority	+0.66
7	Percentage of persons living at a density of over 1½ per room	+0.59
31	Percentage of dwellings which are owner occupied	−0.42
33	Percentage of dwellings which are privately rented	−0.43
4	Percentage of economically active males in social classes I and II	−0.51
38	Percentage of households with seven or more rooms	−0.51
21	Economically active males as a percentage of all males	−0.51
24	Percentage of economically active males in socio-economic groups 1, 2, 3, 4, 13	−0.52
20	Percentage of the total population moving into local authority area during the last five years	−0.14
36	Percentage of persons living at a density of less than 1½ per room	−0.80
41	Persons per room	−0.83

Component 4: 'Ethnic status'	(10.3 per cent explanation)	
Number	Variable Name	Component Loading
16	Percentage of the total population aged 0-4 years	+0.86
12	Percentage of the total population commonwealth born	+0.68
31	Percentage of dwellings which are owner occupied	+0.48
8	Percentage of households lacking a fixed bath	+0.47
13	Percentage of the total population foreign born and born at sea	+0.44

Component 5: 'Migration'		(7.6 per cent explanation)
Number	Variable Name	Component Loading
19	Percentage of total population moving within local authority area during the last five years	+0.82
10	Percentage of dwellings which are multi-dwellings, purpose-built	+0.74
28	Percentage of persons travelling to work by bus	+0.41

Table 6.7: Factorial Ecology of Royal Leamington Spa, Warwickshire: Varimax Component Scores Matrix.

		Components				
		1	2	3	4	5
1.	Enumeration district 01	−0.59	−0.53	−0.61	−0.75	0.14
2.	Enumeration district 02	−1.98	−0.53	−1.76	−0.18	0.44
3.	Enumeration district 03	−0.38	1.55	−0.64	−1.07	1.61
4.	Enumeration district 04	0.01	0.52	−0.16	1.07	−0.37
5.	Enumeration district 05	0.27	2.28	0.36	−1.08	0.54
6.	Enumeration district 06	−1.09	0.62	−0.19	0.03	−0.45
7.	Enumeration district 07	−1.27	−0.28	0.71	0.28	−1.02
8.	Enumeration district 08	−1.56	−1.00	−0.17	−0.56	−0.90
9.	Enumeration district 09	0.61	−0.79	−0.35	−0.10	0.41
10.	Enumeration district 10	−0.68	−0.39	2.54	−0.13	−0.23
11.	Enumeration district 11	−0.17	0.31	1.50	0.20	2.35
12.	Enumeration district 12	0.78	−0.18	−0.81	0.44	−0.32
13.	Enumeration district 13	1.18	0.16	−0.75	0.84	0.92
14.	Enumeration district 14	0.38	0.93	1.24	1.96	−1.07
15.	Enumeration district 15	−0.67	−0.46	−0.46	1.83	0.57
16.	Enumeration district 16	1.09	−1.21	0.49	−1.69	0.04
17.	Enumeration district 17	0.86	−0.77	0.12	0.48	−0.08
18.	Enumeration district 18	1.56	−0.76	−1.04	0.69	0.10
19.	Enumeration district 19	0.97	−1.12	0.74	−1.37	−0.35
20.	Enumeration district 20	0.69	1.69	−0.74	−0.89	−2.32

Positive scores relate to positive loadings, and negative scores to negative loadings, so that maps of the leading scores identify those enumeration districts which are most heavily involved in the 'socio-economic status', 'bed-sitter land', 'housing type and tenure', 'ethnic status' and 'migration' dimensions respectively (Figure 6.6). In a city of only 50,000 population which is divided into a mere 20 zones, sub-areas are unlikely to be extensive or distinctive, but some broad patterns emerge. A basic contrast is drawn between the northern part of Leamington which is characterised by high socio-economic status and high levels of owner-occupation (housing type and tenure), and areas in the south where

Figure 6.6: Factorial Ecology of Royal Leamington Spa, Warwickshire: Distribution of Leading Component Scores.

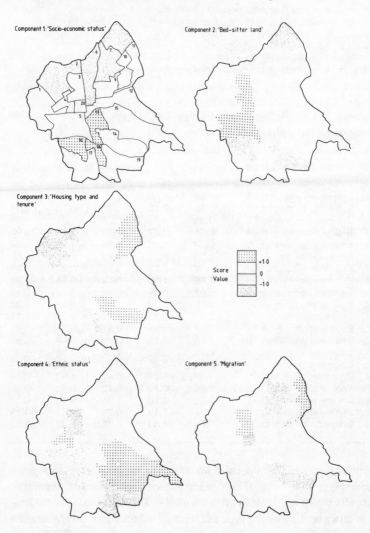

ethnicity and again, housing type and tenure, combine to suggest a different type of social area. A region in the central and eastern part of the city (zones 3, 4, 5 and 20) has a more complex socio-economic profile on account of a combination of migration, ethnicity and bed-sitter land elements. A more detailed insight into internal structure is provided by the factorial ecologies of far larger cities as the work of Robson (1969) on Sunderland, Herbert (1970) on Swansea and Cardiff, Davies and Lewis (1973) on Leicester and Rees (1970) on Chicago has shown. These studies reveal that the overall geographical complexity of the city is in fact the product of the overlap between a number of distinctive, but not necessarily discrete, socio-economic patterns. The involvement of more than one zone in these patterns suggests, moreover, that urban structures are very much more complex, and owe their origins to a more varied set of processes, than those postulated by social areas analysis.

Factor analysis is a complex family of techniques of which the principal components example outlined here is one of the most simple forms. As a consequence of its sophistication, operational decisions must be taken during the course of the analysis involving the selection of input variables, the number of components to abstract and to rotate, and the levels of significance of loadings and scores, so that the object-ivity of the results is open to question. Moreover, there are a number of technical difficulties concerning the distributional characteristics of the data, the size and shape of census districts and the choice of the factor model, which can further distort the analysis as Clark *et al.* (1974) have shown. The large number of studies reported in the literature testifies, however, to the value of factorial ecology as an analytical procedure (Table 6.8). Indeed, an important advantage of the technique is that the initial factor pattern can be subjected to secondary factor analysis so as to produce higher order generalisations about urban social structure.

Johnston (1976) has argued that factorial ecology has advanced the study of urban social geography in a number of ways. By far the most important finding, common to the majority of studies and irrespective of the location and cultural context to the relevant city, is the generality of Shevky and Bell's model of urban structure. Variations in findings may be expected as the data collected by census authorities in different countries are rarely the same, but within these limits, socio-economic status, family status/life cycle and ethnic status consistenly appear among the leading dimensions in factorial analysis. Factorial ecology has, however, taken the understanding of urban social structure further by breaking down the socio-economic characteristics of the city into

Table 6.8: Factorial Ecology and Urban Social and Spatial Structure: Selected Studies.

Country	City	Study
USA	San Francisco	Bell (1955)
	Toledo	Anderson and Bean (1961)
	Washington	Carey *et al.* (1968)
	Akron, Atlanta, Birmingham,	Van Arsdol *et al.* (1958)
	Kansas City, Louisville,	
	Minneapolis, Portland,	
	Providence, Rochester	
	Seattle	Schmid and Tagashira (1964)
	Boston	Sweetzer (1965)
	New York	Carey (1966)
	Newark	Janson (1968)
	Chicago	Rees (1970)
UK	Swansea, Cardiff	Herbert (1970)
	Barry	Giggs (1970)
	Sunderland	Robson (1969)
	Leicester	Davies and Lewis (1973)
	Swansea, Cardiff, Newport	Evans (1973)
	Swansea, Leicester,	Davies (1975)
	Southampton, Colchester,	
	Pontypridd, Llanelli	
	Luton	Timms (1971)
Canada	Toronto	Murdie (1969)
	St Catherines, Trois Rivieres,	Bourne and Barber (1971)
	Sherbrooke, Kingston, Sarnia,	
	Brantford, Niagara Falls,	
	Peterborough	
	Saskatoon, Calgary, Edmonton	Davies and Barrow (1973)
	Edmonton, Quebec City	Bailly and Polèse (1977)
Australia	Brisbane	Timms (1971)
New Zealand	Auckland	Bowman and Hoshing (1971)
		Timms (1971)
	Wellington	McGee (1969)
	Christchurch	Johnston (1973a)
	Auckland, Wellington-Hutt	Johnston (1973b)
	Christchurch, Dunedin	
Finland	Helsinki	Sweetzer (1965)
India	Calcutta	Berry and Rees (1969)

finer detail than that envisaged by Shevky and Bell. For example, many studies identify several related but subtly distinct dimensions of ethnic status. Similarly, family status is frequently divided into two measures —

one concerned with indices of life style, such as dwelling type, dwelling tenure and marital status, and the other with family composition (fertility, working females). Two further features which are common to many studies are the dominant role played by population mobility variables in determining factor structure, and the identification of specialised areas in the city based on a coalescence of socio-economic and ethnic status.

Despite these common elements, factorial ecology studies amount to rather less than a general explanatory statement of urban social structure. Important variations do characterise the social and spatial patterning of cities which, while not negating the validity of the Shevky-Bell dimensions, suggest that they provide only a partial insight. For example, most UK studies identify a sizeable housing type and tenure component which distinguishes the social and economic correlates of owner-occupiers from those who live in public sector housing. Similarly, a 'bed-sitter land' component is normally present which highlights not only the distinctive socio-economic character of a sizeable group in the population, but also their position in the private rented sector of the housing market. While stressing the importance of basic ecological relationships, these observations suggest, however, that the internal structure of the city must be viewed as the outcome of underlying social and economic processes constrained and directed by access to critical commodities like housing and jobs. Locational decisions are not guided solely by social circumstance. Rather they are a product of the inter-actions, and in many cases, conflicts which take place between individuals and those who organise and manage allocation processes in the city.

Conflict/Management Approaches

A focus upon the relationship between individuals and institutions is a recent development in urban geography. In part, it arose out of the growing disillusionment with factorial ecology studies which, despite their increasing technical sophistication, proved incapable of advancing beyond the stage of pattern identification to provide an understanding of the mechanisms and processes which give rise to urban differentiation. A powerful practical stimulus was, however, the downturn in both the UK and US economies in the last decade and its apparent consequences for different parts of the metropolis: peripheral and suburban areas remained relatively prosperous, whereas the problems of the inner city accelerated and compounded. These different spatial effects pointed to the existence of powerful allocative mechanisms within the city which both created and reinforced traditional disparities and injustices. They suggested a situation of conflict rather than concensus, whereby urban structure was the outcome of tensions and frictions inherent in the competition for space in the city. Together, these developments led to a substantive refocusing of urban enquiry. Emphasis in the conflict/management approach is placed upon locational tensions, and upon the roles of those who direct, control and manipulate land uses in the city.

Locational conflict is seen as an inherent consequence of competition for land in the city. For Cox (1973), the city comprises a number of decision making units at a variety of geographical scales. At the micro-level, these might be households or firms: at the macro-scale, they could be branches of local government. Each decision making unit has a set of resources such as personnel, capital and land which it allocates to its activities so as to maximise utility. Thus in fulfilling the needs of its members, a household commits, within a fixed budget, sums and resources to shelter, food, transport and so on. An important characteristic of this allocation process is, however, that individual's utilities are not independent: what is in the interests of some is to the detriment of others. For example, the purchase of a car might bring benefits to one household in terms of improved mobility, but generate pollution and obstruction to the inconvenience and annoyance of another. Two types of externality effect are recognised. The first is 'public behaviour externalities' which cover levels of property maintenance, crime, public comportment and the activities of one's children; the second is 'status externalities' which are those generated by the social and ethnic standing of households. For Cox, such side effects establish a set of attractions and aversions in the city. It is in response to these qualities that urban spatial structure is explained.

Although there is no evidence that social groups view externalities in markedly different ways, the income, cultural and ethnic characteristics of the US population mean that levels of achievement of externalities vary considerably. Cox interprets the emergence of areal divisions within the city as a response to demands for accessibility to those who successfully provide positive externalities, and a demand for physical distance from those perceived to provide negative externalities. Thus location is sought close to groups who maintain the value of their properties, are socially responsible, and who expect and require similar patterns of behaviour in their children, whereas deviant, criminal and low status households are avoided. The emergence of areal patterns in the city is encouraged by the construction of large-scale residential developments which are homogeneous in style, price and quality, and so filter in those who share and are able to maintain and enhance these externalities, and exclude those who cannot. A further factor is the finance and real estate industry which operates a discriminatory system of housing information and allocation as Palm (1976a, b) has shown. The net result is a residential segregation of the North American city based upon such variables as income, social class and ethnic origin.

Once formed, areal divisions in the city are maintained and compounded by intra-zonal flows of finance. Such patterns of exchange have been illustrated in prosaic terms by Bunge (1975) with reference to three interdependent units in Detroit; the city of death, the city of need and the city of superfluity. In this model, low income inner city residents occupy the city of death. They are exploited by the rest of the city because they pay a machine tax to others in the form of job exploitation. The owners of the machines are the wealthy entrepreneurs living in the zone of superfluity on the outskirts who pay them less than their worth for work. In addition, the poor also pay a 'death tax' which takes the form of a surcharge for the higher prices they pay to obtain food, housing, insurance loans, etc. The difficulties of the inner area are exacerbated by lower social service provision and more severe social problems. In contrast, the suburban city of superfluity is the home of the powerful elite class, the professionals, the business people and the politicians. The two areas are separated by the city of need which is 'the home of the white workers, the "hard hats", and the solid union members of Middle America' (p. 158). It is characterised by a net outflow of finance and a deficiency of social provision.

In addition to these essentially private sector flows, the division of the city into local government units, each able to raise taxes and to finance public sector services and provision, generates additional conflicts and disparities. Generally, the capacity of local government to provide public sector services depends upon the tax resources available within its boundaries, but since demand for services is often inversely related to the ability to pay, needs often go unmet in the areas covered by one authority, while elsewhere, tax resources are far from fully exploited. This situation characterises the relationship of central cities and suburbs in the US, and has been described by Cox (1973) as the 'central city — suburban fiscal disparities problem'. It refers to the imbalance between needs for government-provided public services and the tax resources with which to fund them.

Problems of provision arise because central areas generate high levels of demand for public services, but have only a low yield tax base. This situation is largely a result of the concentration there of low income groups, ethnic minorities, crime, an ageing population and fire prone housing, and is increased by migration and commuting trends. Thus central areas have a tendency to lose high value, high tax paying businesses and activities, and to replace them by low income, low yield uses. Similarly, more daily movements take place by suburban residents to the centre than of central residents to the suburbs, and this imposes a further burden of tax demands on the central city relative to its tax capacity.

Fiscal collapse in decaying central areas is a general characteristic of US cities as Alcady and Mermelstein (1977), Hill (1977) and Tabb (1977) have shown. It is especially pronounced in older metropolises such as New York where the central area has to carry the additional burden of supporting nationally prestigious services such as libraries, zoos, centres for the performing arts, and many of the social costs associated with the headquarters offices of major corporations. Declining central areas induce additional frictions into the city which widen the gap between areas which offer positive and negative externalities. They reinforce mechanisms of attraction and avoidance which, according to the locational conflict approach, give rise to areal differences in the city.

Within the same general context of locational conflict analysis, research workers in the UK have concentrated their attention on the ways in which allocation mechanisms in the city are 'managed' by 'urban gatekeepers'. The main reasons for this slightly different approach lie in the social and political characteristics of US and UK cities. One difference is that cities in the UK are generally (but with the exception of London), under the jurisdiction of a single local authority which is responsible for the provision of most public services. This means that at least in theory, centre-periphery imbalances in tax yield and public provision should be more easily resolved. A second contrast is the more homogeneous and stable character of UK society which leads to less extreme pressures of inter-group attraction and aversion. A third is the very different pattern of housing tenure with regard to public sector involvement: 30 per cent of UK housing is public sector as compared to four per cent in the US. Housing and property development in fact have so far constituted the central focus of interest in UK managerial studies. The aim is to examine the extent to which urban managers, a group which for Pahl (1975) includes 'public housing managers, estate agents, local government officers, property developers, representatives of building societies and insurance companies, youth employment officers, social workers, and magistrates and councillors' exert an influence on the allocation of scarce resources which may reinforce, reflect or reduce social inequalities in the city (Table 6.9).

Table 6.9: The 'Management' of the City: Selected Studies.

Focus	Studies
Building societies/estate agents	Ford (1975), Lambert (1976), Grigsby (1970), Weir (1976), Duncan (1976), Williams (1976, 1978), Boddy (1976, 1980)
Real estate agents	Palm (1976a, b, 1979), Wolfe *et al.* (1980)
Public housing provision	Paris (1974), Lambert (1976), Gray (1976), Dennis (1978), Bird (1976), Harloe *et al.* (1974), Karn (1976), Paris and Lambert (1979)

Underlying the managerial approach is a simple model of the UK housing/property development system which identifies groups of actors or agents, and financial and decision making links between them (Figure 6.7). Members of the public, in their varying roles as householders, students, hospital patients, workers, etc. generate a wide ranging set of demands for built space which the housing/property development system seeks to satisfy either through the creation of new or improved space, or by the exchange of existing properties. Unlike most commodities, property has a very high initial cost relative to the users' income and so requires an elaborate means of funding. This role is performed by finance capital institutions which, with the exception of overseas borrowing, draw the majority of their funds from the savings, pension contributions and insurance premiums of individuals. Unlike industry, which seeks to make profits from the sale of manufactured goods, finance capital institutions increase the value of sums deposited with them by receiving interest on investments. Collectively, funding property construction or sales is only one of a number of investment options open to finance capital institutions, although it is the primary function of building societies and property companies. The building and construction industry uses finance capital to create the built environment. Again, the exercise of a number of development decisions determines the relative volumes of different types of property which are constructed or which change hands. Valuation, selling and establishment of title are interrelated activities undertaken by members of the chartered surveyors, estate agents and legal professions. They close the demand-supply cycle

Figure 6.7: The Finance Capital and Property Development System in the United Kingdom.

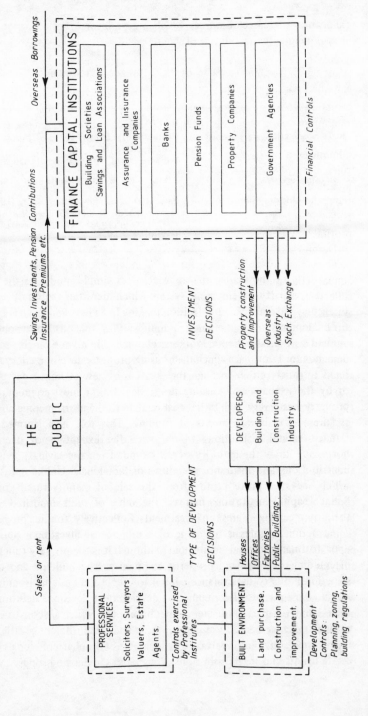

by arranging the transfer of property between vendors and purchasers.

The housing/property development system is regulated at various points by external agencies. Finance capital institutions are subject to control by the Bank of England which establishes rules of conduct, liquidity levels (the ratio of loans to deposits) and interest rates. Construction activities are subject to building regulations which lay down minimum design and construction standards, and to planning controls which determine what can be built where. Both sets of regulations are administered by local government, but within the broad framework of rules and policies established by central government. The activities of architects, surveyors, valuers, estate agents and solicitors are overseen by their professional institutes. Members are required to comply with codes of practice designed to establish the basis for a professional and profitable working relationship with their clients.

Despite the obvious complexity of the housing/property development system, it seems reasonable to assume that the net product of annual construction and sales will be more or less a reflection of collective social needs. A high level of response may be expected of a system in which everyone involved seeks to satisfy expressed or antici-pated demands for built space. Analysis of its operations, however, suggests that the system has a tendency to work to the advantage of some groups and to the detriment of others, thereby generating and widening social divisions in the city. These dysfunctions are illustrated with reference to three areas of decision making within the system: type-of-development decisions, mortgage allocation decisions and improvement grant decisions.

Conflicts which are associated with types-of-development decisions were most clearly evident in the central areas of the major UK cities, and especially London during the 1960s and early 1970s. This was a period of major urban renewal in which large amounts of finance capital were devoted to central area reconstruction. Major investment in housing, shops, schools and other public buildings was urgently required, but a disproportionate share was channelled into highly profitable but far less socially desirable office development. In a detailed analysis, Ambrose and Colenutt (1975) have shown that the process was managed by a small group of estate agents and property developers who, together with their engineers and architects, amassed enormous personal fortunes and indeed public esteem through their property dealings. They were assisted considerably by a set of fiscal circumstances which resulted in their firms paying very little taxes on profits, and by local authority planners who either co-operated, or were powerless to prevent large-scale office development in their areas. On the debit side, office construction led to the demolition of many houses and shops, to job losses, and to the destruction of long established and thriving inner city communities. Moreover, by buying up streets of old but structurally sound properties in preparation for demolition and office building, the developers cast the shadow of development and planning 'blight' across large areas of the inner city.

The financing of housing purchases is a second area of decision making which can have major social and spatial implications for the city. Unlike office development which is largely funded by property bond companies, banks and pension funds, 80 per cent of all housing finance in the UK is made available by building societies. In assessing submissions for a loan, societies take into account the income, security of income and the general creditworthiness of the applicant. Although manual workers can achieve relatively high wages in the early stages of their work careers, they have less chance of increasing their income through time and are, therefore, less favoured for mortgages than white collar and professional groups who have greater levels of job security and an incremental salary structure. Similarly, old and non-standard dwellings such as multi-occupied dwellings and converted houses are seen as a bad risk. These lending policies combine to produce a distinctive spatial pattern of mortgage flows within the city. They lubricate the suburbanisation process by diverting funds to new housing on the expanding urban fringe and discriminate against lower income house-holders seeking to purchase cheaper properties in the inner city. Parts of the inner city may in fact be totally excluded from consideration for mortgages through a practice that has been termed 'redlining' (Karn, 1976, Lambert, 1976, Williams, 1978). This means that societies refuse to lend on properties, which may be acceptable in themselves, if they are located in declining areas where property values are likely to depreciate.

Despite the difficulties of attracting private sector finance to the inner city, public sector aid in the form of housing and area improvement grants is normally available. For example, local authorities in the UK are empowered to designate General Improvement Areas in which individual house improvements are backed by environmental improvements carried out by the local authority, and high stress Housing Action Areas in which greater direct assistance for repairs and modernisation is made available to landlords and owner-occupiers. Studies of the allocation of grants, however, have revealed cases where it was not the areas and the households in greatest need who were given the greatest assistance. For example, in the declaration of GIAs, Duncan (1974) has shown how local authorities have veered away from areas with high proportions of coloured households and of the very worst housing with the aim of selecting those areas most likely to show to best effect the impact of improvement. Similar findings are reported for Bristol by Bassett and Hauser (1976) and Bassett and Short (1978). Rather than significantly ameliorate the situation, public sector interventions, which by their very nature are selective, would seem to be insufficient to compensate for the powerful private sector forces which create and maintain concentrations of low quality housing and high social stress in inner city areas.

The conflict/management approach is a comparatively new departure in urban spatial analysis, and so the overall perspective is far more complete. There are many points of decision making within the finance/ property development system that have yet to be investigated by urban geographers. Rather than contributing a general theory of locational conflict and managerial intervention, it provides at present an overall framework within which to identify those whose decisions determine the shape of the city and to examine their effects. Though different in emphasis, locational conflict and managerial studies have a common basis. Both are concerned with frictions and allocative mechanisms within the existing urban economy. It is with a challenge to the basic economic and social formation, however, that Marxist explanations of urban structure begin.

Marxist Analysis of the City

Marxist perspectives represent a radical departure from the ecological, trade-off, social area/factorial ecology and conflict/management approaches to the analysis of urban structure. Though focusing upon different social and economic processes, most non-Marxist studies share a common approach. They begin with implicit assumptions of, and an acceptance of, a capitalist economy, and seek explanations of urban structure in terms of the processes and relationships that are themselves a product of that type of economy. The Marxist approach is altogether more fundamental. In place of secondary processes and their outcomes it questions the underlying social and economic order. Social and spatial divisions in the city are seen as both an inevitable and a necessary consequence of a capitalist mode of production.

Despite his enormous output, Marx himself wrote very little on the city. He was primarily concerned with economic structure, economic changes, social class formation and struggles in society as a whole. It has been for subsequent theorists, especially Harvey (1973), to extend Marx's analysis into an urban context. The theory outlines the mechanisms through which land in the city is allocated to individuals on the basis of their incomes. It further describes the ways in which ownership determines rent which in turn determines land use structure (Table 6.10).

Table 6.10: Basic Lines of Argument in a Marxist Theory of Urban Land Use.

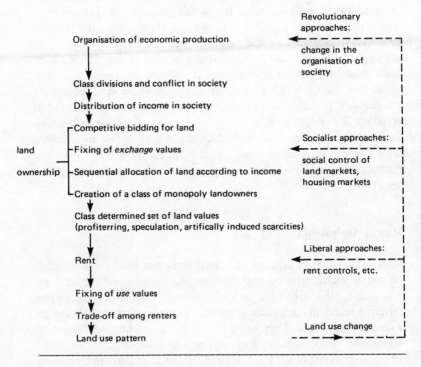

Basic to the Marxist approach is an analysis of the organisation of production. In every society except the most primitive, the bulk of the population is engaged directly in production while a tiny minority controls their labour and the things they produce. For example, in medieval society the slave owner or the feudal lord contolled the labour and product of the slave or serf. In modern free enterprise economies, the owners of industry and commerce — the capitalists — direct the labour of wage earners and own the fruits of their labour. This relation-ship between a ruling class and subservient group is one of opposition and conflict. Class division and friction is, therefore, a primary stimulus for social change.

Since capitalists make decisions based upon their desire to make profits, they try in every way possible to pay workers only part of the value their labour produces; an amount just sufficient to sustain life and to reproduce more workers for the next generation. Marx called the value produced by workers, in excess of what they are given, surplus value. It is the basis for profit and property income in society. In capitalism, where the class struggle permeates every segment of society, some sectors of the working class find themselves in a position in relation to their employers that is more favourable than other sectors of the working class. This allows the advantaged sectors greater success in their struggles for higher wages and better working conditions. More-over, employers also have an interest in paying their workers a range of wages. These wage differentials, by dividing workers and granting nominal rewards for those working within the system, militate against the formation of a strong class consciousness and working-class militancy. The capitalist organisation of the economy thus gives rise to a highly skewed income distribution as a consequence of major differences in levels of reward between the ruling and working class, and more minor wage variations within the working class.

What are the urban geographical consequences of a class-divided, income-differentiated society? In *Social Justice and the City,* Harvey (1973) addressed this question by applying Marx's economic analysis to urban land use. As the land available in the city is limited in quantity, it has economic value. For most commodities, economists distinguish two measures of value. The first is use value and is a measure of worth derived from putting the commodity to work; the second is exchange value and is the value realised when the commodity is sold. A simple example of use value is the enjoyment and benefit derived from riding a bicycle; its exchange value is the price it fetches when traded on the open market. Two characteristics, however, mean that use value and exchange value have special significance when applied to land. One is that land changes hands very infrequently and so purchase at one point in time enables the use of land to be enjoyed, or to be leased or rented to others, for long periods of time. The other is that land is fixed and immobile so that the purchase of land in effect bestows monopoly rights on the landowner. Land (and improvements on land in the form of buildings) is a commodity that individuals cannot do without. It follows, therefore, that those who own land are placed in a powerful economic position over those who do not.

Both the exchange value of land and land ownership are determined by competitive bidding. Harvey envisages land parcels in the city being allocated in very much the same way as seats are occupied in an empty theatre: the first person who enters has n choices, the second, n-1 choices, the third, n-2 choices, and so on. If those who enter do so in order of their bidding power, then those with the highest incomes have the greatest choice while the low income take up whatever space is left when everyone else has exercised choice. The low income may indeed find that all the land is spoken for so they will be forced to rent. The most important characteristic of this sequential occupance process is, however, that the relative cost of space in the city is low for high income groups and is high for low income groups. The richest have to pay but a small amount more than the next richest group in order to purchase the prime locations in the city. The poorest, conversely, have to extend their finances to the limit to buy into the market at the lowest level.

The ability to raise a sum of money equivalent to the exchange value (i.e. the purchase price) of land leads to the creation of a class of monopoly landowners in the city. The purchasing power of high income groups is such that large tracts of the city are typically owned by a small number of landowners who effectively control the urban land market. Some plots will be occupied by the landowners themselves, but others will be rented or leased to those who are unable to purchase. Marxist analysis suggests that landowners in a capitalist system have a vested interest in maximising the return or profit on their land investments so they set rents as high as possible and reduce expenditure on repairs and maintenance to a minimum. It may indeed be in their interests to allow their properties to become derelict if this enhances the prospects of comprehensive redevelopment involving high rent-yielding office space. Similarly, properties may be left unlet if this creates a shortage which increases rents elsewhere. (This strategy incidentally explains one of the major paradoxes of the urban housing market, which is that the inner area is commonly that part of the city with the most severe housing shortages and problems of homelessness, and yet at the same time it has the largest number of vacant and empty properties.)

It is by fixing and manipulating rents that the landlord monopoly class dictates the urban land use pattern. Rent, for the tenant, is the price that is paid to occupy space in the city and it therefore represents the use value of land. If rents are fixed at a high level, then the land will be restricted to high quality retail and commercial uses. Similarly, if rent levels are low, residential uses may become established. This explanation of the origins and role of rent is important, and constitutes the most fundamental difference between non-Marxist and Marxist analysis of the city. Non-Marxists see the use of land as determining its rent, whereas Marxists argue that it is rent as socially prescribed and manipulated which determines use. In competing for leasing space in the city, industrialists and householders place bids in relation to the rent structure. The outcome, as the trade-off model predicts, is an orderly, commonly concentric land-use pattern.

Some important practical implications follow from a Marxist analysis of urban structure. Capitalism, it is argued, produces wealth for those who own the means of production but also exists to maintain security. This means that areas of decay and decline in the city are both an inevitable and a necessary consequence of a capitalist mode of production: inevitable because profit-seeking will always drive up rents to a level where they detract from other items of household expenditure; and necessary because without such exploitation, the property owning/rent fixing class would lose their advantaged position. A parallel point was made in the Community Development Project which was concerned with inner area economic problems in Coventry (CDP, 1975). The authors argued that the urban economy both generated and required a pool of low cost labour which could be employed and laid off at short notice so as to cushion capitalist industry against fluctuations in the economic cycle. Deprivation and poverty according to this interpretation are not reflections of personal or institutional failures; rather they are inherent and inevitable consequences of the capitalist organisation of society.

The logic of these arguments from an urban policy point of view is that state intervention in the city in the form of rent or wage controls is inadequate since these are directed essentially at symptoms rather than causes. To remove social and spatial injustices in the city it is necessary to change the economic organisation of society which gives rise to them. For example, if public ownership replaced private ownership of land, then space could be allocated so as to satisfy social needs. For Harvey (1973), the justification for regarding housing as a social (as opposed to a private) good is that all housing built before, say 1940, has effectively been paid for. The only costs attached to it are maintenance and service charges. 'We have an enormous quantity of social capital locked up in the housing stock, but in a private market system for land and housing, the value of the housing is not always measured in terms of its use as shelter or residence, but in terms of the amount received in market exchange which may be affected by external factors such as speculation. In many inner city areas at the present time, houses patently possess little or no exchange value. As a consequence, we are throwing away use value because we cannot establish exchange values. This waste would not occur under a socialised housing market system, and is one of the costs we bear for clinging tenaciously to the notion of private property' (p. 138).

Harvey's analysis paints a utopian picture of the type of society and city that can be created by changing the organisation of the means of production. It suggests that collectivisation and state control of land and property will remove basic conflicts between groups and so lead to a system which is both socially and spatially just. Although the internal logic is impressive, questions must however be raised about whether the outcomes which are envisaged in theory can be achieved in reality. One difficulty with this type of assessment is that the centrally planned economies of the Eastern Bloc, which might provide some support for Harvey's hypotheses, are all comparatively recent, twentieth century, indeed mostly post-war, creations. They may not have existed long enough for the social pattern to be radically transformed. Another is the practical problem of measuring social inequalities in countries in which official pronouncements, and hence official statistics, declare that none exist. That the major cities of the Eastern Bloc appear to exhibit many of the social and spatial characteristics, and certainly some of the social problems of their Western counterparts, suggests however that centralised control may produce outcomes which are similar to those which are commonly associated with conditions of private ownership and capitalism. For Pahl (1979), all societies, whether capitalist or centrally planned have social injustices and areas of deprivation which are a reflection of the distribution and availability of scarce resources. Attempts to reduce these inequalities produce complex systems of bureaucracy which generate their own socio-political consequences. Resources are allocated in all societies through a system of managerial bargaining. It is in the rules, procedures and preferences which these organisations create, rather than in the underlying socio-political structure, that explanations of urban social and spatial patterns lie.

The Marxist approach has been criticised on account of its ideological content. It is held to be subjective and non-scientific because it is grounded upon an alternative and a non-conventional view of society. Moreover, by seeking to promote fundamental social and spatial changes it is held to be 'political' and so worthy of lesser consideration than *status quo* models which, it is argued, are value-free. Although these criticisms have been levelled specifically at Marxism, they are in fact of general applicability and relevance. All theories and models in the social sciences, both Marxist and non-Marxist, are ideological because all are based ultimately upon certain assumptions concerning the underlying organisation of society. By being concerned with private land and business ownership, and a competitive free market economy, even central place theory, Weber's industrial location model and the bid-rent model of urban land use presuppose a capitalist mode of production and organisation of the economy and so are inherently ideological. A second observation is that those who seek to retain the *status quo* are as 'political' as those who attempt to change it. Politics refers not merely to the activities of radicals and revolutionaries but also to those who use power to maintain and enhance established positions and privileges. What distinguishes Marxist approaches is that the ideological and political content is overt and explicit rather than covert and hidden. This certainly means that they are more susceptible to criticism and rejection on ideological grounds, because those grounds are clearer, but not that they are any less useful, or less worthy of consideration as a body of theory. In any assessment of theories of urban structure it is important to be aware of the ideological base, whether those theories are Marxist or non-Marxist in origin.

Marxist analysis has far-reaching implications for the study of the city. It raises questions of a more fundamental nature than those traditionally addressed by urban geographers. In essence, the city is seen as a microcosm of society as a whole, and it is at the societal rather than at the urban level that explanations of city structure are developed. It is, moreover, at the societal level that solutions to urban problems must be sought (Hamnett, 1979; Short, 1977). Marxist analysis is logical, consistent and persuasive but the ideological content which characterises all models of urban structure means that a wholly objective and dispassionate evaluation is difficult. It is for individual urban geographers to decide whether the class conflict model as advanced by Marx corresponds with their own views on the nature of social organisation.

Conclusions

This chapter has outlined the major approaches which urban geographers have adopted in analysing the internal structure of the city. The very different perspectives involved reflect both the highly complex nature of urban social and spatial structure, which necessitates the consideration of a wide range of viewpoints, and the influence upon urban geography of changing interests and emphasis in geography as a whole. Despite the considerable contemporary concern with conflict/management and Marxist analysis, it is important not to dismiss preceding viewpoints out of hand. Though somewhat simplistic, the ecological approach generated some profound insights and raised questions of sufficient importance to stimulate social area and factorial ecology approaches. Indeed, it provided a highly appropriate framework for discussing and analysing the social and spatial characteristics of the rapidly-growing nineteenth and early twentieth century city. Similarly, although perhaps more applicable at a secondary level, the relevance of bid-rent and trade-off mechanisms has been broadly confirmed by Marxist analysis. No single approach would seem capable of accounting for all the social and economic processes that differentiate cities. Explanations of urban structure require a synthesis and fusion of insights drawn from a wide variety of perspectives.

Irrespective of their contrasting philosophical and ideological standpoints, the different approaches to urban structure all emphasise the close link between socio-economic process and urban pattern. Such a relationship is in one respect obvious for in countries in which eight out of every ten people live in urban areas, the city *is* society and distinctions between them are meaningless. In another respect, however, this observation is of profound importance since it underlines the basic rationale for the existence of urban geography as a specialist field of study. This book began with the assertion that urban geography is one of a number of complementary subject areas in the social and environmental sciences which focus upon the city. It is by analysing and explaining the spatial consequences of social and economic processes, at the intra- and inter-urban scales, that urban geography contributes to the broad multi-disciplinary understanding of the contemporary city and urban society.

7 URBAN PLANNING AND URBAN POLICY

The preceding chapters have focused upon the processes of growth and location which primarily determine the distribution and organisation of cities and city systems. Historically, the operation of these forces was largely free from any form of public accountability so that urban development took place in an unregulated and uncontrolled manner. At the same time, though cities emerged as centres of wealth and prosperity they were also characterised by severe overcrowding, high levels of morbidity and mortality and chronic deprivation and poverty. Today, an elaborate and powerful system of planning exists in the UK and to a lesser extent in North America, which aims to circumscribe urban development and to direct it towards socially desirable goals. Urban geographers, in seeking to identify and account for the spatial characteristics of towns and cities can no longer restrict themselves to a consideration of underlying economic, social and environmental processes: the effects of planning upon urban patterns and problems must be considered as well.

For urban planning to exist, there must be a broad consensus among the population in a country that serious problems affect their cities and that these can best be tackled through government intervention. This in turn requires a willingness on the part of individuals to relinquish some of their 'rights' to property and development which they enjoy in a free market situation and to accept the principle that land use should be centrally regulated or controlled in the public good. For urban planning to be successful, there must be further agreement as to its objectives and mechanisms, that is, what its goals are and how it sets out to achieve them. Recognition of the need for urban planning emerged in the UK, Europe and North America during the late nineteenth and early twentieth centuries as a response to the perceived problems of the industrial city. It took the form of a broad-based intellectual and social movement headed by a small but influential group of politicians, philosophers and idealists who sought to reform the city. But despite the existence of very similar urban problems, and exposure to the same arguments and advocacy for planning, national responses were very different. For example, electorates in the UK progressively demanded, and have accepted the implications of, a powerful and all-embracing system of urban planning whereas state intervention in the US city amounts to little more than zoning and minimum land use control.

Rather than a uniform pattern, the extent, objectives and machinery of urban planning vary markedly from country to country.

A second important characteristic is that urban planning has evolved considerably over time in response to the changing nature of urban problems. At the turn of the century, attention was primarily focused upon overcrowding and health, and controls were placed upon building and land use in a belief that improvements to the physical environment would alleviate the major social problems of cities. Within the last two decades, however, there has been a growing appreciation of the severity and persistence of urban social and economic problems, especially in inner areas, and an awareness that the control of land use and the lay-out of settlements are by themselves insufficient. This in turn has led to an extension of the basic regulatory approach to include a consideration of all those mechanisms and processes which determine urban form. In addition to physical design, intervention in a wide range of urban processes is now practised in many countries within the context of general urban planning. Although it commonly incorporates a physical element, general planning represents an altogether more comprehensive approach to urban problems than design planning and involves strategies concerned with employment, housing, transportation and service provision. It is undertaken by land use planners working in conjunction with specialists in these fields. Planning has become essentially a process of general policy formulation involving the identification of goals for urban development, the specification of the steps necessary to reach them and the monitoring and evaluation of the pace of achievement. The city is seen as a product of many individual but interrelated processes each of which can be appraised, regulated and monitored so as to achieve general urban objectives.

Different national experiences and changes of objective and structure over time preclude any general assessment of the impact of planning on the city. Following an historical outline of the origins of urban planning, this chapter therefore, focuses upon the particular examples of urban planning in the UK and USA. Some of the dilemmas and choices facing modern cities, and the people who plan them, are finally illustrated by detailed reference to the inner city problem in the UK.

The Origins of Urban Planning

Urban planning has emerged over the last century as a response to the manifest and well documented problems of the industrial metropolis.

It is debatable whether living conditions were significantly worse in the nineteenth century city than they had been in the countryside prior to industrialisation, but the concentration of deprivation in the urban slums which made poverty and disease visible and even threatening to the middle and upper classes, led to these conditions being defined as a basic problem for society. Two types of reaction to this situation were present at the time. One, represented by Marx and Engels, was revolutionary and proposed the overthrow of the social and political system that created Disraeli's 'two nations' of rich and poor; the conservative alternative involved the basic acceptance of the urban-industrial system but the use of state intervention to ameliorate its worst excesses. It was the latter argument, articulated in the UK by the factory and sanitary reformers and reinforced by the success of several early housing and new towns schemes, that led to the emergence of modern urban planning.

That planning in the sense of designing new communities offered a means of escape from the problems of the nineteenth century city was shown by the Utopian Socialists, a term coined by Marx to describe a group of thinkers who sought to improve working class conditions by individual benevolence, philanthropy and enterprise. First and perhaps foremost among the group was Robert Owen who proposed, in *A New View of Society* (1813), the creation of agricultural villages of between 800 and 1,200 people, catering for all the social, educational and employment needs of the community. His plans, submitted to a parliamentary committee looking at problems of the working class in 1817, envisaged residence for members of these communities close to their place of employment, communal living for the older children, central heating, private lodging only for families with children under three years old, and an agricultural base but with some industry on the outskirts. Owen himself set the example by establishing an industrial complex at New Lanark, Scotland, which provided excellent working and living conditions, cheap but subsidised shops and an adult education unit. It pointed the way to the ideal community, free from ignorance, vice and disease and providing good housing and a range of social facilities for its residents.

Owen's educational and social reform measures, including his emphasis upon 'the formation of character', encouraged many subsequent idealists. The actual design of a town to accommodate communities of a large size was suggested in mid-century by Buckingham (1849) who advocated a Model Town Association and described the layout of an ideal settlement, Victoria, for 10,000 people. This particular scheme was never carried out, but utopian communities on a somewhat less grandiose scale were started in the 1850s by Titus Salt at Saltaire, Bradford and by Price's Patent Candle Company at Bromborough Pool, Cheshire, while the company town of Pullman, constructed near Chicago in the 1880s, is an example of a similar development outside the UK. The century ended with the building of three further experimental communities by Lever, at Port Sunlight, Liverpool (in 1888), by Cadbury, at Bournville, Birmingham (in 1879) and by Rowntree, at Earswick, York (in 1904), the first two of which contributed markedly to the growth of the planning movement. Port Sunlight was distinguished by particularly attractive housing, and a low level of crime and disorder in the community which suggested that the major social evils could be overcome by intelligent physical design. Similarly, the success of Bournville, which housed both company and non-company employees in a spaciously laid out and imaginatively landscaped suburban setting, underlined the merits of social balance and high environmental amenity. Together, these utopian communities demonstrated that major improvements to the social conditions of the working class could be achieved by rehousing them in carefully designed, planned communities.

A second and parallel stimulus to planning, which emphasised the need for direct state intervention in the urban environment, was represented by the urban Sanitary Reform movement. The health of the urban population was an increasing cause for concern in the mid-nineteenth century as a succession of epidemics ravaged the densely populated areas of the major British cities. By simple mapping of the place of residence of the victims, Dr J. Snow the local general practitioner was able to domonstrate that the cause of the 1848 cholera outbreak in Soho, London, was contaminated drinking water (Stamp, 1964). This evidence of a deficiency in public utilities reinforced the argument for government action and under pressure from enlightened politicians such as Shaftesbury, Torrens, Cross and Chadwick, a series of legislative measures, which established basic levels of sanitary provision and standards of building, was enacted. The 1875 Public Health Act consolidated previous measures and introduced a set of codes or requirements in respect of the level, width and construction of new streets; the structure of houses, their walls, foundations, roofs and chimneys; the layout of buildings and external space requirements, and their sanitary facilities; as well as providing powers for closing down dwellings which were unfit for human habitation. These measures did much to reduce levels of morbidity and early mortality in the city, although they did nothing to promote equity and increase social opportunity.

The idealism of the Utopian Socialists was mirrored at the turn of the century by the ideas of Ebenezer Howard as set out in his important and influential work, *Tomorrow: A Peaceful Path to Real Reform* (1898). For Howard, the twin evils of society at the time were the simultaneous depopulation of the countryside, which he saw as man's natural habitat, and the overcrowding in new and rapidly expanding industrial cities. Both town and country acted like magnets, attempting to draw people towards them, but as each presents advantages and drawbacks, a compromise location, which he termed town-country, was preferred (Figure 7.1). Upon this argument, Howard based his proposals for garden city communities which he portrayed on a grand scale. Each garden city was to have a population of 32,000 and Howard envisaged that they would be developed in groups of six around a comprehensively planned central city of 58,000 people to make totally planned urban units of a quarter-of-a-million people. The city would be self-contained in terms of employment, possessing its own industry, commerce, shops and agricultural production, indeed all the facilities required by the population. The most important practical effects of Howard's book, which was republished in 1902 as *Garden Cities of Tomorrow,* was the founding in 1899 of an Association to promote the garden city ideal. Two garden cities, Letchworth and Welwyn, were built by the Association in 1901 and 1920 respectively, with the latter attempting for the first time to secure a 'social balance' by skilful integration of varying socio-economic groups. The Garden City Association was a major stimulus to the formation in 1914 of the Town Planning Institute which as the Royal Town Planning Institute is today the major organisation for professional planners in the UK.

In parallel to the development of utopian idealism in Britain, alternative and indeed more ambitious designs for the city were being advocated in Europe and North America. By far the most important focus for forward thinkers on the continent was the Italian Futurist movement launched in a manifesto by Marinetti in 1909, which embraced a wide range of artists, musicians, sculptors and philosophers, anxious to break with the past and to experiment with, and create, new and dynamic modes of expression. The New City was one of the main ingredients of this movement, and was regarded as symbolic of a new world which would be comprehensively planned on a grand scale. Ideas were presented by architects such as Sant Elia in a series of futuristic urban designs which featured high rise towers and elevated roadways which bore little resemblance to the building patterns and street plans of the city at the time. Emphasis in these designs was placed upon zoning and the physical separation of industrial, commercial, residential and recreational areas.

Figure 7.1: Howard's Three Magnets.

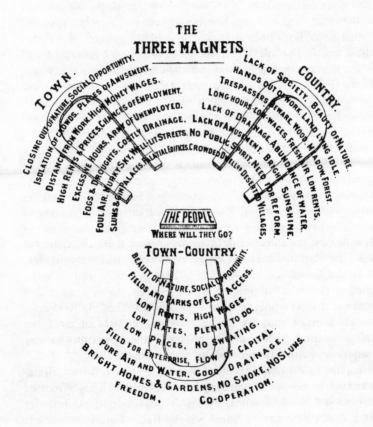

Source: Howard (1898), p. 46. Reprinted by permission of Faber and Faber Ltd. (1946).

Construction would be undertaken with the mass production methods that were being introduced into industry, and using the new and exciting materials of glass and concrete. These proposals were drawn together in *Towards a New Architecture* (1925) by the Swiss architect, Le Corbusier. In his City of Tomorrow, designed for three million people, there would be subways for servicing, a co-ordinated transport system, high rise business and entertainment centres surrounded by five to seven storey residential blocks, and detached dormitory garden suburbs. Changes on this scale amounted not merely to totally new cities, but required new mass production societies to live in them.

Although few of Le Corbusier's plans were implemented in their entirety, his work had a profound effect upon the subsequent development of planning and upon the shape of cities throughout the world. As the major visionary of twentieth-century architecture, his ideas influenced a generation of disciples from whose ranks were drawn many of the leading city planners of the immediate post-war period. Particularly important was Le Corbusier's conception of the city, indeed the house, as a machine and his plans for enormous residential and office blocks, which with their open promenade decks and exercise areas, incorporated many of the constructional and design features of luxury ocean going liners. Similarly, high rise buildings separated by public open space were preferred to individual houses with their own gardens. Changes in the urban fabric of this magnitude necessitated wholesale rather than merely piecemeal development. The opportunities provided by post-war reconstruction enabled many of these schemes to be put into practice with important consequences for the architectural character, layout and skyline of many European cities.

Reactions to the nineteenth-century city in North America were altogether more diverse. For Berry (1973), the preconditions for urban planning in the USA were laid eighty years ago by the Progressive intellectuals, a loose knit group of sociologists, economists and political scientists including White, Howe, Addams, Follett, Dewey, Royce, Giddings, Cooley and Park who, in examining the shortcomings of American capitalism, recognised a clear case for public intervention in, but not control of, the economy. Among arguments for government regulation of business, employment and politics, they advocated the appointment of specially trained experts who would administer the cities. Against this background, designers considered ways of improving the appearance and quality of life in American cities. In common with Ebenezer Howard's ideas, landscape architects such as Olmstead, Davis and Vaux produced designs of residential areas as 'cities in a garden', which led to the building of Llewellyn Park, New Jersey, Riverside, Illinois and Brookline, Massachusetts as America's first planned romantic suburbs. More ambitious designs for changing the appearance of the American city were published after the Chicago Exhibition of 1893 as part of the City Beautiful Movement (Wilson, 1980). The basic idea was for the planned unity of the entire city as a work of art, supported by a master plan for land use, and by comprehensive zoning ordinances to maintain that plan. New environmental ideals were advanced, as was the notion that planning was essential if the industrial city was to be saved from what was seen to be progressive physical and moral deterioration. The City Beautiful Movement provided a major stimulus for American city planning such that by 1913, 43 cities had prepared master plans and 233 were engaged in some form of major civic improvement (Berry, 1973, p. 22).

Despite their very different complexions, a belief that social conditions were a product of the physical environment was common to these various social and intellectual movements. It was because people were forced to live in squalid and overcrowded conditions that cities were characterised by disease, crime and deprivation. The answer was to sweep away the cities of the past and to replace them by new planned communities. This argument led in turn to visions of a utopian society characterised by stability, health and affluence that would live in the planned city. Inspired by this philosophy, the planning proposals which emerged in the immediate post-war years were heavily biased towards physical design. Primary emphasis was placed upon controlling land use as a means of achieving, indirectly, a set of loosely defined and highly idealistic social objectives.

Urban Planning in the UK: the 1947 Planning System

Although urban planning in the UK was inspired by reformist reactions to the nineteenth-century city and the physical solutions which followed from them, its cause was advanced in the 1930s by the growing focus of attention upon population distribution and land use issues. As London and the cities of the West Midlands, Yorkshire and Lancashire continued to grow, so succeeding censuses, mapped by Fawcett (1932) and Taylor (1938), pointed to an increasing concentration of the population in a narrow axial belt which extended diagonally across England from south east to north west. In recognising this pattern, politicians and academics expressed concern at its long-term implications for the balance of regional opportunity, especially as it was the areas outside this belt that were characterised by high unemployment and general industrial decline. An additional factor in population distribution was the spread of the built-up area, made possible by suburban train and bus services. This further increased the possibility that the major cities would coalesce into one vast urban region covering the central area of England and it was to inquire into the causes and implications of these develop-ments that the Royal Commission on the Geographical Distribution of the Industrial Population, popularly known as the Barlow Commission, was established in 1937. A related fear expressed in 1925, by the newly-formed Council for the Preservation of Rural England, was that cities were so encroaching upon good agricultural land that this was seriously undermining the food producing capabilities of the nation. This argument, supported in the 1930s by evidence produced in Stamp's Land Utilisation Survey of Great Britain, led to the work of the Scott Committee on Land Use in Rural Areas which reported in 1942. In the late 1930s, a number of quite different philosophical and practical arguments pointed to the need for greater government control of land use and urban development. Nineteenth-century reformist zeal, utopian idealism and the concern for population distribution and land protection were the major strands which inspired the establishment of the 1947 planning system in the UK.

In legislative terms, the framework of contemporary planning in the UK was created by a series of measures enacted by Parliament in the immediate post-war years (Table 7.1). The Distribution of Industry Act of 1945 provided the foundation of regional planning in the UK by designating development areas in which grants and aid to industry were available. In the New Towns Act of 1946, provision was made for scheduling sites for new towns and for setting up Development Cor-porations to build and manage them. The Town & Country Planning Act of 1947 was a comprehensive measure which created the structure, machinery and indeed the profession of urban planning. On these bases emerged what has come to be known as the 1947 planning system.

Among the most important features of the system were those intro-duced by the Town & Country Planning Act (1947). Prior to this legislation, the approach to urban planning was largely piecemeal and unco-ordinated, a product in large part of the division of local responsi-bilities for planning amoung 1,441 local authorities in Enland and Wales. By 1942, 73 per cent of land in England and 36 per cent in Wales was subject to interim development controls but only five per cent and one per cent respectively was actually subject to operative schemes (Cherry,

Table 7.1: The Origins of the 1947 Planning System in England and Wales: Major Reports and Legislation, 1940-7.

1940	Report of the Royal Commission on the Distribution of the Industrial Population (The Barlow Report, Cmd.6153)
1942	Report of the Expert Committee on Compensation and Betterment (The Uthwatt Report, Cmd.6386)
1942	Report of the Committee on Land Utilisation in Rural Areas (The Scott Report, Cmd.6378)
1943	The Ministry of Town & Country Planning Act
1943	The Town & Country Planning (Interim Development) Act
1944	The Town & Country Planning Act
1945	The Distribution of Industry Act
1945	Abercrombie's Greater London Development Plan
1946	The New Towns Act
1947	The Town & Country Planning Act

1974, p. 100). After 1947, responsibility for urban planning in England and Wales was vested in the 145 County and County Borough Councils which were required to prepare and submit development plans for their area to the Ministry of Town & Country Planning, itself created in 1942. These plans indicated how land in the area was to be used, and were subject to revision every five years. The basic principle enshrined in the Act was that of private land ownership but public accountability in use, so that owners of land were not allowed to undertake development without first obtaining permission from the local planning authority. Appeal against a local planning authority's decision lay to the Minister. A second principle embodied in the Act was that of community rather than individual gain from land betterment, so that when land was developed, the increase in its value which resulted from the granting of planning permission, was reserved for the community by the imposition of a 100 per cent land development tax. This provision was however removed in 1952. Powers were also given to local authorities to acquire land for public works and housing projects and for this and similar purposes a compensation fund was established.

Important changes to the process of plan making were introduced in the 1968 and 1971 Town & Country Planning Acts. In place of the single rigid development plan with its specific land use focus and five year life expectancy, these acts created a two tier system of planning at the structure (or strategic) and local scales. Structure plans are comprehensive documents which seek to translate and apply national and regional economic and social policies to the area for which the local authority has planning responsibility. As such, they attempt to combine considerations of a wide range of social, economic and physical objectives for the area within a broad strategic framework. More detailed structure plans may be provided for individual towns and cities in the area. Local plans seek to apply the strategy of structure plans to particular areas and issues and to make detailed provisions for development control. Local plans may be produced for Districts, when comprehensive planning of large areas is required, for Action Areas, when major improvement, development or redevelopment is scheduled, or for particular subjects within the structure plan. The broad philosophy behind these changes is that of providing a comprehensive but flexible response to the changing needs of the area.

For Hall *et al.* (1973), the 1947 planning system in England and Wales, reflecting the various philosophies, ideals and movements which led to its creation, was geared to the achievements of a number of different objectives at a number of different spatial scales (Table 7.2).

Table 7.2: Objectives of the 1947 Planning System in England and Wales.

Scale	Primary Objectives	Secondary Objectives
National-regional	Maintain the existing regional balance	
Sub-regional or city regional	Urban containment; protection of the countryside and of natural resources; creation of self contained and balanced communities	Prevention of scattered development; building up of strong service centres
Local	Maintain accessibility to urban functions; maintain and promote a high level of physical and social environment	

Source: Hall *et al.*, 1973, p. 39.

At the national-regional scale, the idea of maintaining existing regional balances applied most directly to regional planning. The most important focus for the urban geographer is, however, upon the criteria which operated at the sub-regional or city regional scale and which were essentially concerned with distributions, activities and movements in a small region. Here Hall *et al.* identify three important and interrelated objectives: urban containment — the most important, protection of the countryside and the creation of self-contained and balanced communities. Two other objectives at this scale were subsidiary and were the prevention of scattered development and the building up of strong service centres. Finally, at the local scale, there were two objectives — enhancement of accessibility to urban functions and the promotion of a high level of physical and social environment.

Other than issues of regional balance at the national-regional scale, these objectives were broadly advanced through the development plan/development control process. In drawing up development plans, local authority planners took account of the pressures for expansion in the cities in the area, the quality and availability of land, population distribution and the range and accessibility of services, and devised land use categories accordingly. Despite their different emphasis, many of the planning objectives were complementary and self-reinforcing. For example, a green belt designated to contain urban expansion would also protect the countryside, maintain and enhance the physical environment and prevent scattered development. Similarly, urban growth could be equated with the ideals of self-containment and social balance by the construction of new towns. Emphasis in planning between 1947 and 1968 was, however, placed heavily upon design. Attention was directed primarily towards the distribution and appearance of the built-up area rather than intervention in the underlying social and environmental processes which were responsible for urban developments and distributions.

In parallel to development planning undertaken by local authorities, the urban objectives of the 1947 planning system were pursued by central government through the new towns policy. Although basically inspired by the idealism and experiences of the early garden city movement, new town thinking in the post-war years owed much to the Barlow Report and to Abercrombie's *Greater London Plan,* published in 1945. The former, in analysing the geographical distribution of the industrial population, underlined a general need for satellite towns to prevent overconcentration, while the latter provided more detailed arguments with reference to the population and housing problems of London. The New Towns Act of 1946 was concerned to promote new towns in furtherance of a policy of planned decentralisation from congested urban areas. It placed great weight on attaining a satisfactory social balance of population in freestanding and self-contained communities.

Table 7.3: British New Towns: Progress and Potential as at 31 December 1976.

	Date of Designation	Area in Hectares	Pop. at Designation	Pop. in 1976	Pop. Target in 1976
First generation (Mark I) new towns					
Stevenage	Nov. 1946	2,532	6,700	74,000	105,000
Crawley	Jan. 1947	2,396	9,100	75,000	85,000
Hemel Hempstead	Feb. 1947	2,391	21,000	78,000	85,000
Harlow	March 1947	2,588	4,500	81,000	90,000
Aycliffe	April 1947	1,254	60	26,000	45,000
East Kilbride	May 1947	4,148	2,400	76,200	90,000
Peterlee	March 1948	1,133	200	27,500	30,000
Hatfield	May 1948	947	8,500	26,000	29,000
Welwyn	May 1948	1,747	18,500	41,000	50,000
Glenrothes	June 1948	2,333	1,100	33,700	70,000
Basildon	Jan. 1949	3,165	25,000	91,890	130,000
Bracknell	June 1949	1,337	5,149	45,000	60,000
Cwmbran	Nov. 1949	1,278	12,000	45,000	55,000
Corby	April 1950	1,791	15,700	53,500	70,000
Second generation (Mark II) new towns					
Cumbernauld	Dec. 1955	3,152	3,000	45,000	100,000
Skelmersdale	Oct. 1961	1,669	10,000	41,000	80,000
Livingston	April 1962	2,708	2,000	29,000	100,000
Redditch	April 1964	2,906	32,000	53,200	90,000
Runcorn	April 1964	2,930	28,500	54,600	100,000
Washington	July 1964	2,271	20,000	46,000	80,000
Irvine	Nov. 1966	5,022	34,600	52,305	120,000
Milton Keynes	Jan. 1967	8,900	40,000	77,000	250,000
Peterborough	July 1967	6,453	81,000	109,000	180,000
Newtown	Dec. 1967	606	5,500	7,700	13,000
Northampton	Feb. 1968	8,080	133,000	158,000	230,000
Warrington	April 1968	7,535	122,300	135,400	201,500
Telford	Dec. 1968	7,790	70,000	99,700	220,000
Central Lancashire New Town	March 1970	14,267	234,500	248,000	420,000

Source: adapted from Blake (1977), pp. 89-103.

The 28 new towns which have been created under the Act were designated in two major phases (Table 7.3). Of the 14 first generation new towns, the eight in the outer London ring, together with East Kilbride outside Glasgow, were intended to advance the specific objectives advocated by Howard, Barlow and Abercrombie, of dispersing population from overcrowded urban areas (Figure 7.2). Included in this list is Welwyn Garden City which was redesignated as a new town. The remaining five towns in this group were designated to aid specific regional or industrial development policies in the development areas. They included Aycliffe, to house workers on an adjoining trading estate, Peterlee, to provide a central point among East Durham mining villages, Cwmbran, to serve the industry of South East Wales, Corby, to house steel workers and Glenrothes, to serve the East Fife coalfield.

Figure 7.2: British New Towns.

Unlike their garden city predecessors, the Mark I new towns aimed to achieve a social balance in both individual neighbourhood units and in the overall community. For example, Crawley strived to reflect the class character and social balance that existed in the country as a whole. Thus, when there was found to be 20 per cent middle class 'nationally', plans were made for the integration of 20 per cent middle class in the neighbourhood units as well as in the town at large (Ratcliffe, 1974, p. 45). The concept of 'neighbourhood' was itself an important element in new town design and housing developments were in many cases grouped into neighbourhood units, each provided with a comprehensive range of community facilities. A second feature of the Mark Is was their target populations which ranged from 10,000 (Aycliffe) to 80,000 (Hemel Hempstead). With these comparatively small sizes, it proved difficult to achieve the planned social and educational balance and the new towns provided little in the way of direct stimulus to local economic development.

The second phase of new town construction was begun in 1956 with the designation of Cumbernauld. In social terms this represented a movement away from seeking a balance by complete integration and its community structure was far less physically determined than the first generation of new towns. There was less adherence to the formal idea of an architectually designed neighbourhood unit and less physical division of the town. Indeed, by allowing for higher densities and pedestrian and vehicle separation, it was hoped that higher levels of social interaction would be achieved. Cumbernauld was joined after 1959 by 13 further Mark II new towns (Table 7.3). With the single exception of Newtown, designed to help solve the problems of the thinly populated hill area of mid-Wales, these were intended to accommodate conurbation overspill. They were therefore located close to the major centres of population in the Greater London, West Midlands, Merseyside, Manchester and Tyneside areas. In all these later Mark II new towns, increased mobility and physical accessibility have tended to invalidate even further the idea of a fixed residential neighbourhood size. Moreover, the idea of separating functions such as employment, shopping, education and residence has also been questioned and in Washington, for example, a basic village unit of about 4,500 has been created with its own jobs and shops. Schools, pubs and community meeting places are also provided in the local 'village' centre. The unit is designed to be of a size suitable for peoples' daily needs and is self-contained to the extent that residents have a sense of belonging. In part, different social and spatial designs were necessary because of the increased target populations fixed for the Mark IIs. With anticipated sizes of 250,000 and 420,000 respectively, Milton Keynes and the Central Lancashire New Town have the effective status of new cities (Table 7.3).

The new towns programme has been supplemented in England and Wales by the expanding towns scheme. Introduced by the Town Development Act of 1952 this has sought to cope with urban growth by providing financial support and encouragement to local authorities to agree among themselves regarding the planning and management of overspill schemes — nearly 70 expanding towns arrangements were nominated under the programme between 1952 and 1970 with a combined targeted population of 50,000. This compares with an estiamted increase of three-quarters of a million people over the base population in the 28 places designated as new towns.

For Hall *et al.* (1973), the most important achievement of the 1947 planning system in England and Wales is that megalopolis has been denied. Post-war urban growth has been contained in both population and geographical terms to such an extent that the coalescence of adjacent cities has been thwarted and valuable agricultural land has been protected. Green belts around the conurbations and major free-standing cities have stopped peripheral extension and the effect of development control in county and rural areas has been to concentrate new development in existing towns and villages. The new towns, by syphoning off population from the conurbations are an important part of this achievement and as financially viable, socially stable and econ-omically prosperous communities, rank as major successes in their own right. As a consequence of planning, the impact of urban growth in terms of rural land conversion and environmental destruction has most probably been less severe than would have been the result of urban patterns which might have arisen as a result of *laissez-faire* processes. Two other consequences have been less welcome. The first, which is associated with the lack of effective local controls on the location of employment, has been the growing separation of residental areas from main centres of work and services, and this has resulted in a major suburbanisation of the urban population. The second has been soaring inflation in land prices which planning has in part created and has certainly been powerless to stop. On balance, however, the 1947 planning system appears to have been broadly successful in achieving the objectives which were set for it at its inception.

Planning the American City

Although urban planning in the USA shares the same reformist roots as in the UK, its evolution and contemporary structure are very different. The most fundamental contrast, reflecting the greater emphasis placed upon private enterprise, is the limited degree of government intervention in the US city and the extent to which it seeks to promote private as opposed to community ideals. A second characteristic is that American planning is not obligatory, as in the UK and this, together with the fragmented structure of local government, means that the content of planning is both local and varies from place to place. A tradition of national concern for cities was however established in the 1930s which, in providing direct aid for urban assistance, has only recently been paralleled in the UK. There is thus no system of planning in the sense of a common framework with a clearly defined set of physical, social and economic objectives: land use in American cities is controlled at the municipal level by zoning regulations while specific urban problems, such as housing, are addressed through federal financial and aid policies.

Comprehensive zoning ordinances were first introduced at the turn of the century as a means of promoting the ordered urban structures that were idealised in the City Beautiful movement. Their purpose was to promote the health, safety, moral and general welfare of the community. These they sought to achieve by preventing the worst effects of uncontrolled urban and industrial development by establishing standards of compatibility, density, light and space. In prohibiting uses which would be harmful to other property, zoning ordinances operate in an essentially negative direction. Equally, by excluding detrimental developments, they serve to enhance the established uses in the area. For Delafons (1969), American land use controls are designed to preserve the interests of existing property owners rather than to promote desirable patterns of land use in cities. As such, they are devices for protection rather than for planning.

Every municipality has the power to control land use and private development and as the existence of control may be a consideration for receiving federal aid, most cities now have a system of zoning ordinances approved by the city council. 'Planning' as opposed to zoning is a separate activity, traditionally undertaken by a city planning commission, which has the responsibility of preparing a 'Master Plan'. This is not so much a blueprint for land use or the solution of urban problems, as much as a broad picture of how the city might be improved through a programme of public works. Planning responsibilities at the municipal level are therefore limited in nature. While seeking to promote developments which are both profitable for the individual and the community, they aim to mitigate the adverse effects of growth and development.

At the national scale, the concept of planning was central to the liberal philosophy which inspired Roosevelt's New Deal in the 1930s. It took the form of national and regional initiatives designed to ameliorate the worst effects of the Depression. Two aspects of New Deal policies had particular implications for the city. The first was the creation of the National Resources Planning Board to promote planning initiatives at the state level, many of which were directed towards solving urban problems. Indeed, the Board's two major publications, *Our Cities: Their Role in the National Economy* (1937) and *Urban Planning and Land Policies* (1939) together comprised as much of a national urban policy as the US ever had (Catanese, 1979, p. 23). The second feature was the programme for the construction of greenbelt towns undertaken by the Rural Resettlement Administration. Three thousand were originally envisaged but only three, Greenbelt Maryland, Greenhills Ohio and Greendale Wisconsin were ever built. In design terms these were superficially analogous to Howard's Garden Cities but as dormitory suburbs of Washington, Cincinnati and Milwaukee respectively, they lacked the self-sufficiency and social balance that characterised the British new towns.

In the immediate post-war period, such planning as was undertaken by the federal government was primarily oriented towards recovery and the development of the US economy. The most important urban programmes were those concerned with public housing, assistance and aid as introduced under the National Housing Acts of 1949 and 1954. These strengthened and extended the slum clearance and urban renewal projects and encouraged cities to resolve their own problems with federal financial assistance. To ensure that urban renewal and housing schemes were well thought out, the government offered aid through the Section 701 Programme which provided matching grants to communities which produced comprehensive plans. Although this arrangement channelled aid directly to the depressed areas of the major cities, the emphasis which was placed upon physical redevelopment meant that comparatively little was accomplished in social and economic terms. Moreover, as the number of programmes proliferated, so the problems of co-ordination and implementation increased. These particular deficiencies were redressed somewhat during the 1960s by the War on Poverty and the Model Cities Programmes. Whereas the former was a general attack on poverty, and so applied to both urban and rural areas, the latter, as its name implies, was specifically urban in direction. The Model Cities Programme was designed to encourage participating cities to develop a concerted offensive against social and economic problems as well as physical decay. Accordingly, it required the co-ordinated efforts of all relevant agencies. In essence, eligible cities received one-year planning grants with which to prepare Comprehensive Plans to 'improve the quality of life' in their Model Neighbourhoods. Both implementation and ongoing planning would occur over a five year demonstration period, during which time funding would be available through appropriate federal aid programmes. Although the concerted approach was attractive in theory, the programme's basic and broadly stated objectives, which included co-ordinating and concentrating federal, state and local resources, developing innovative initiatives and involving local residents in planning, were difficult for the cities to define in terms that were locally relevant and achievable. Moreover, it proved to be highly sensitive in political terms and received varied support from successive administrations. The War on Poverty and Model Cities Programmes were eventually replaced during the 1970s by systems of revenue sharing through which cities were given blocks of financial aid according to a needs formula. These community development block grants were to be used in whatever way the city concerned considered to be the most appropriate. Inevitably, this raised and indeed continues to raise fundamental questions about whether planning and aid should be directed towards physical redevelopment or social and economic regeneration.

Despite basic differences in philosophy and structure, two threads are common to the development of urban planning in the UK and USA. The first is the initial diagnosis of urban problems as being primarily environmental and the consequent emphasis in early planning which was placed upon improving the physical layout and design of cities. The second is the more recent recognition that the problems of the contemporary city, far from being physical, are social and economic in origin. This change in thinking is obviously critical since it conditions the whole planning response. Instead of simple land use controls there is a need for a more fundamental approach which seeks to regulate and change the mechanisms which determine the distribution of social and economic activity in the city. One consequence is the eclipse of land use planning by comprehensive urban policies governing the range of financial and human resources in the city. The other is the widening of choices which the increased scope of state intervention in the city introduces, together with the associated questions concerning who decides, who loses, and who benefits from planning. As it is a political activity, such issues are of course present in all forms of planning, but they assume overt and critical significance when governement policy becomes directly and closely concerned with the provision and quality of homes, jobs, services and life chances in the city. The dilemmas this raises for the city are fundamental and far-reaching. They are well illustrated by the debates generated and range of strategies which have been proposed to reverse the decline of inner urban areas.

Inner City Problems

That an inner city problem of major proportions exists in many UK cities was acknowledged in 1977 by the publication of the government White Paper *Policy for the Inner Cities* (HMSO, 1977). This document recognised that inner area decline and deprivation had largely replaced sprawl, overspill and inter-regional imbalances (the problems of the inter-war and immediate post-war years) as the most serious difficulties facing UK cities. In drawing upon the experiences of research and previous policy initiatives, the White Paper identified the major economic, social and environmental components of the problem and proposed a course of remedial action that forms the basis of present inner city policy. Although there is substantial agreement concerning the dimensions of the problem, the factors responsible are, however, much disputed so that inner area decline is a central focus of academic, political and doctrinal debate. The range of alternative viewpoints indeed represents a major obstacle to the development of successful planning policies for the inner city.

Four basic components to the inner city problem were identified in the 1977 White Paper. The first was economic decline associated with the contracting industrial base of inner areas and its implications for employment. In part, this problem is long-standing and is associated with the general decentralisation of population and industry as major UK cities have recently, in geographical terms, turned themselves 'inside out' (Gripaios, 1977; Bull, 1978). An important contributory factor is the run-down of traditional inner city services and activities, both private and public sector, including docks, railways, distribution, warehousing, printing and publishing (Department of the Environment, 1977a, b). Economic decline in the inner city is accentuated more generally, however, by the recession in the UK economy which began in the early 1970s and which is responsible for the closure of many activities in the manufacturing sector (Lloyd and Mason, 1978). Particularly important in this context is the shut-down and rationalisation of branch plants, often of multinational corporations, and the associated 'ripple effects' which result in the failure of many small dependent firms (CDP, 1975). Despite the high accessibility to markets which results from central location, inner areas suffer a combination of disadvantages which include relatively high costs of industrial land and rates (local taxes), shortage of suitable industrial premises, problems of inadequate access and limited opportunities for expansion, and labour constraints including a shortage of female workers. As a consequence there has been a failure to attract new industry so that inner areas are characterised by severe unemployment. This is most serious for those who fall into one or more of the following categories of being black, disabled, elderly, semi-skilled or unskilled, and a school leaver.

A second feature of many inner city areas is the condition of the physical environment which in general is characterised by decay, deterioration and the lack of amenities. Such conditions arise basically because of the age of the inner areas of UK cities, most of which were built in the mid and late nineteenth century, but which have not benefited from the continuing investment and improvement which has been made to the nearby retail and commercial areas of the central business district. Environmental decay has, however, been exacerbated in many cities by shortcomings in the development process whereby land has been cleared and then left idle for long periods. Similarly, the publication of long-term proposals for comprehensive redevelopment or road widening naturally discourages private investment and may cast the shadow of planning blight across a wide area of the inner city. An additional factor has been the stop-go nature of public sector financing which has affected the continuity of many local authority development programmes.

Of particular concern in inner areas is the condition of housing which because of age and neglect may lack adequate damp proofing, insulation, modern plumbing and wiring and amenities such as a fixed bath, hot water and an inside WC (Figure 7.3). The lack of housing improvement is related to tenure as inner areas typically contain large numbers of properties which are let privately on a furnished or unfurnished basis (Figure 7.3). Absence of household amenities does not necessarily, however, imply a need for demolition and wholesale redevelopment, for many houses are structurally sound and can be refurbished and brought up to date at comparatively low cost so as to

Figure 7.3: Selected Socio-economic Characteristics of Coventry, 1971.

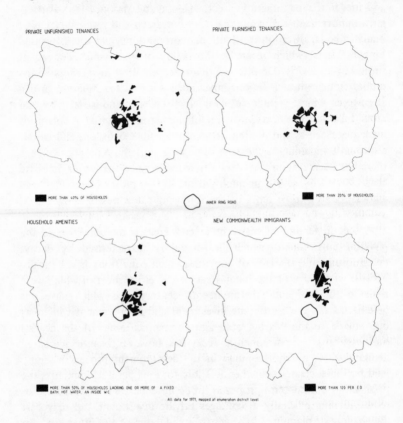

provide cheap and attractive accommodation.

Social disadvantage was identified by the 1977 White Paper as the third major component in the inner city problem. It characterises those who are poor, as well as many of the infirm, elderly and ethnic minority groups. Social disadvantage is partly a product of high unemployment, but the level of earnings in inner areas also tends to be low because of the nature of available work. In addition, a sizeable proportion of the inner city population is outside the labour force because of age or infirmity. The persistence of pockets of social disadvantage in inner areas lends support to arguments that deprivation is cyclical and tends to be transmitted from one generation to the next. Those with few job skills have little purchasing power and can only afford cheap and inferior housing in inner city areas with a poor physical environment and inadequate social and educational provision. These conditions tend to produce or exacerbate stress in the home so that children perform badly at school, and failing to gain the qualifications necessary to compete successfully in the job market, the cycle is completed and begins again. A second feature of social disadvantage is that it is collective and affects all the residents, even though, individually, the majority of the people may have satisfactory homes and worthwhile jobs. It arises from a pervasive sense of decay and neglect which affects the whole area through the decline of community spirit and through greater exposure to crime and vandalism.

Although their presence within the urban population does not of itself constitute a 'problem', ethnic minorities and immigrant groups tend to concentrate in parts of the inner city (Figure 7.3). This introduces community and race relations tensions and adds a fourth dimension to the underlying economic and social difficulties of the area. Ethnic minorities in inner areas suffer the same kinds of disadvantage and deprivation which are experienced by all those who live there, but as a visually and culturally identifiable group, they are most open to discrimination in the job and housing markets. They are, moreover, easy targets for those who seek simple causes and innocent scapegoats for the country's or the city's decline. As well as general difficulties, a combination of language problems and poor educational provision commonly gives rise to particular problems at school which means that many immigrant children fail to gain the qualifications they need for employment. In contrast to economic decline, physical decay and social disadvantage, which are common to most large cities in the UK, the ethnic minority problem has however a regional incidence. Sizeable coloured populations live in many of the large cities of the south, midlands and north-west, whereas many Scottish and north-eastern cities have relatively few members of ethnic minorities among their numbers (Peach, 1975; Jones, 1979). There are on the other hand, sizeable ethnic minority communities in the outer parts of some cities and in many smaller towns and cities such as Bradford, Blackburn, Leicester, Wolverhampton, Bedford and Slough. A further complication is presented by the different ethnic mix of cities. There are important differences among immigrant groups, and especially between West Indians and Asians, in terms of community and family support, level of acculturation and assimilation, and linguistic and educationsl achievements.

The problems experienced in inner city areas are complex. Many can be attributed to industrial decline and unemployment which are major causes of poverty and low incomes, but personal factors which may condemn individuals to hardship and stress, such as old age, infirmity, loss of marriage partner or ethnicity, are clearly independent of the economic climate. Similarly, the general environmental conditions in terms of soundness and quality of buildings and range of neighbourhood amenities reflects the age of the area and levels of investment over the years rather than the prevailing financial and political circumstances. Despite the existence of separate strands, the complex dynamics of disadvantage and deprivation tend to combine and compound these difficulties so that economic hardship, personal stress and poor physical environment become coincidental and self-reinforcing. That these problems are most apparent in inner areas does not, however, mean that they are exclusive to those areas or that the underlying causes are necessarily geographical. Society's failings can be reflected in space just as much as spatial problems infringe on society.

Strategies for the Inner City

For Kirby (1978), four general strategies may be identified among the many palliatives which have been proposed for the inner city (Table 7.4). The first, as outlined in the 1977 White Paper *Policy for the Inner Cities* (HMSO, 1977) seeks to revitalise inner areas through a programme of economic assistance and co-ordinated aid by central government in association and partnership with local authorities. The second would involve a relaxation of the housing market in order to allow households greater movement within and between sectors and areas and in consequence greater social and geographical mobility. A means of dissolving the problem by moving industry and population to peripheral urban locations and new towns forms the basis of a third strategy. Finally, major social and economic changes, both reformist and revolutionary, are advocated by the left in a belief that inner city problems stem from injustices and divisions which are inherent in capitalist society. These strategies reflect fundamentally different conceptualisations of causes especially the extent to which they are spatial or structural in nature. They are concerned, respectively, to defeat, disperse, decentralise and deny the inner city problem.

Present policy for the inner city stems from a diagnosis that a decline in economic fortunes lies at the heart of the problem. It seeks, primarily, to strengthen inner area economies by making financial assistance available to industry as well as helping to promote social and

Table 7.4: Inner Cities: Alternative Strategies for Improvement and Re-vitalisation.

Explanation of Problem	Primary Objective of Strategy	Status of Strategy	Example of Provisions
Industrial decline/employment loss	Economic revitalisation	Present policy	Industrial Improvement Area Programmes. The Urban Programme. Enterprise zones. Partnership schemes
Problems arising from individual psychological and social handicaps and inadequacies transmitted from one generation to the next	More integrated and self supporing individuals and families	Present policy	Compensatory social work, support and self help
Inflexibilities in the housing market	Improved geographical and social mobility through reform of the housing market	Proposed	Changes to the system of housing finance and public sector housing allocation
Obsolete physical environment and consequential social problems	Relocation of population to new planned communities	Historic: implicit in garden city and new town policies, 1903-1970	New town construction
Maldistribution of resources and opportunities	Redistribution of resources and opportunity through taxation and public spending	Recently and presently implemented but with a very low level of commitment	Positive discrimination in social service provision. Redistributive taxation
Problems arising from failures of management and administration	More total and co-ordinated approach by bureaucracy	Implemented in part, but with very low level of commitment	Rational social planning. Area management

Explanation of Problem	Primary Objective of Strategy	Status of Strategy	Example of Provisions
Problems arising from the divisions necessary to maintain an economic system based on private profit	Changes in political conciousness and organisation	Projected	Fundamental redistribution of political power and control

environmental improvements. Four separate programmes within the strategy have a specific industrial emphasis. The first is the scheme for distributing aid through the Industrial Improvement Areas in which authorities are able to provide advances and rehabilitated factories to attract migratory industry. The second is the designation of enterprise zones as free trade areas in which public controls are held to a minimum so as to encourage the establishment and growth of industry. For example, industry within these areas is exempted from certain taxes, planning regulations and bureaucratic intervention. The objectives of the Urban Programme are more general in that assistance is made available to cover the costs of industrial, environmental and recreational provision. It represents a means whereby assistance can be directed via the local authorities to voluntary bodies and ethnic minorities. The fourth approach involves the designation of partnership schemes whereby central government can work with and aid those local authorities where deprivation is thought to be particularly serious. Assistance in these cases is provided to support an inner area programme which is drawn up by central government and the partnership authority. These individual initiatives, as well as Comprehensive Community Programmes and policies on population movement, form a complex package of intervention and assistance which is designed to arrest and reverse the economic, social and environmental decline of inner urban areas. Underlying the whole approach, however, is an assumption that inner city problems are spatial rather than structural. The strategy is strongly geographical in that government intervention is directed towards particular parts of the city rather than at the basic social and economic divisions which many argue lie at the heart of the inner city problems.

While accepting the need for economic recovery, the *Inner Area Studies: Summary of Consultants' Final Reports* (Department of the Environment 1977b) identified a basic lack of residential and hence social mobility in the contemporary British city. Particular criticism was directed at the operation of the housing market in inner areas which, by drawing in the deprived and disadvantaged and restricting the outward movement of the upward mobile, it was argued, created and compounded social problems in the inner city. A strategy follows from this analysis which would introduce changes designed to improve the housing situation of those who wish to remain in inner areas and to allow those who wish to leave to do so. Central to the approach would be an increase in the size of the local authority housing sector, the extension of criteria of eligibility for public housing to include unemployment and family size and structure, and measures designed to facilitate the movement of council tenants both within and between local authority areas. It would also involve the introduction of changes in the lending practices of building societies which would prevent the savings of inner area investors flowing out to finance suburban house purchases. The most important criticism of this approach is that it is highly specific, and questions inevitably arise as to whether housing reform by itself can be sufficiently powerful to promote the necessary degree of social revival in inner areas. Housing improvements alone would seem to be insufficient if employment opportunities and educational, recreational and service provision remain unchanged.

The third strategy for the inner city identified in Kirby's (1978) analysis is that of decentralisation. It involves evacuation and the relocation of inner city industry and population in new planned communities. The strongest arguments in favour of this approach are the achievements of the post-war new towns, and especially their success in attracting industry and jobs which suggests that they could provide important new opportunities for re-located inner city residents. Construction, population growth and investment could also be used as in the case of the Mark II new towns, to stimulate the declining regional economies thereby solving two major planning problems with a single planning strategy. The decentralisation argument is, however, based upon a supposition that inner city problems are inherently locational whereas if they are structural, far from being resolved, they will simply be transferred to the new towns. Moreover, the addition of displaced inner city populations is likely to disturb the social balance of existing new town communities. Although strongly supported by the building and construction industry, the level of government spending that would be necessary suggests that the decentralisation strategy is unlikely to prove attractive to governments committed to monetarist economic policies. It would in any case involve an unacceptable loss of existing infrastructure and investment in the inner city and as such leaves questions about the future role of those areas unanswered.

The most radical approach is to question the existence of an inner city problem and to argue that what exists in inner areas is a set of social and economic problems which are expressed in spatial terms. Inner areas according to this analysis are not discrete entities with their own distinctive characteristics, rather they are part of, and so reflect, the tensions and injustices prevalent in the national social and economic system. A central observation is that unemployment, poverty, deprivation and disadvantage characterise the population at large and are not limited to those who live in inner areas. Inner city problems cannot, therefore, be divorced from broader economic, social and political forces: they are inherently structural rather than spatial in character. Two different strategies follow from this analysis, differing in degree but not direction (Lawless, 1981). The first is reformist but not revolutionary and advocates a reorganised and much increased public sector involvement in inner city community aid and employment provision. The second is radical and in rejecting the social and economic *status quo* advocates a fundamental change of the political system that would give political power to disadvantaged groups in the inner city and elsewhere and so dramatically improve their life chances.

For the reformists, changes in the organisation and activities of government are seen to be necessary so as to stimulate employment increase and community revitalisation in inner areas. Underlying this argument is a belief that inner area employment decline, while related to the contraction of manufacturing industry, is compounded by the rationalisation and run down of public sector services and nationalised industries, especially in the transport and related sectors, which are traditionally strong in inner areas. In addition, regional policy by encouraging the transfer of capital and industry to the assisted areas and to new and expanded towns has added further to the residual pool of unemployed or at best, low paid workers, in inner areas. These trends can be reversed, it is argued, by vastly increased public expenditure so as to stimulate employment in inner areas. This would take the form of both expenditure to increase the number of public sector jobs directly, and incentives designed to encourage and, some would advocate, compel, private industry to locate in inner areas. Although such industrial regeneration is essential, many reformists question its sufficiency and argue the case for parallel changes in the scale and organisation of welfare aid to the inner city. Criticisms here are levelled at the fragmented nature of existing public sector approaches which are characterised by divisions of responsibility for social, health, housing and environmental services both between and within central and local government. In part, these shortcomings have been recognised by the designation of partnerships which are designed to facilitate co-ordination between central and local government on specific projects, but reformists call further for a 'total approach' by which government aid is extended to cover the range of social, economic and environmental problems of inner areas. The reformist position on government initiatives also tends to stress the need for intervention at the local level through area management which is seen as a means of encouraging, among other benefits, better service delivery, improved two-way information contacts between government and governed, more resources for the deprived and great responsibility for resources by local residents.

Although these proposals are altogether more far-reaching than those which form the basis of the present economic revitalisation approach to inner cities, the likelihood that they will promote the necessary degree of structural change may still be questioned. They are, for example, castigated by radicals as measures designed merely to reduce unemployment and dissent to levels at which they cease to be a threat to the ruling class. For the revolutionary left, inner city problems are a reflection of basic injustices and inequalities that are inherent and inevitable in capitalist society. They can be overcome not by reversing particular spatial grievances or by obtaining additional resources for certain areas but only by fundamentally changing that society. This process can, it is argued, best be achieved by raising the political consciousness of deprived residents and by organising and uniting isolated minority groups into a force for community action. If collectivation were indeed to prove possible and national working class combines were to emerge, the revolutionary left would hope to attain certain objectives. Amongst the most important of these would be to control investment in the community interest, to produce a socially efficient distribution of industry, and to ensure that the costs of social change are shared evenly by all sections of society. Two observations, however, indicate that this sequence of action and change is most unlikely. The first is the extent of differences and divisions within the working class which, combined with a generally low level of political consciousness, suggests that the prospect of collective action of a revolutionary nature is remote. The second is that inner city dissent is as likely to be stirred up and directed by the far right, against targets such as coloured immigrants, as by the far left, against the corporate state. The complex and varied nature of local problems means in fact that inner cities are a primary focus of attention for factions representative of all shades of political persuasion and opinion. Under these circumstances it seems most probable that middle ground compromise rather than extremist policies will receive greatest support.

The likelihood of arriving at a generally acceptable approach which can command wide support is increased by the fact that many of the individual urban initiatives are not incompatible, and can therefore be applied in combination. For example, policies of economic revitalisation do not necessarily exclude new town construction altogether. Indeed, a selective relocation of the unemployed from inner areas to new planned communities could reduce the overhead social costs in the inner city and so give it more chance to prosper. Similarly, although they are commonly advocated by supporters of very different ideological persuasion, policies of social support and compensatory social work in inner areas have for many years been paralleled and reinforced by policies of positive discrimination in education and health service provision. As well as choice of overall strategy, decisions must be made regarding the combination of individual initiatives within the strategy, and the emphasis and level of resources each will receive. Viewed in these terms, a major attraction to planners of broad-based centrist policies is that individual initiatives can be given differential weighting at short notice so as to suit the needs of prevailing political opinion.

The strategies which have been advanced to assist inner areas highlight the dilemmas which face contemporary cities and the people who plan them. They reflect differing shades of opinion as to what are the causes of inner urban decline and how they can best be reversed. That planning appears to be at a crossroads may be an illusion, for the containment, protection of countryside and new towns policies of the immediate post-war era did not go unchallenged at the time. The scope and power of general urban planning is, however, such that the opportunity for intervention in the city, and hence the range of choices for urban change, is far wider than when urban planning was narrowly concerned with land use and physical design issues. Planning is no longer restricted to outcomes in terms of patterns and distributions, rather it seeks to regulate the underlying socio-economic processes. A second contrast is the entrenched ideological positions which underpin these different planning strategies. As a basically political activity, all planning decisions are ideologically flavoured whether they are concerned with maintaining the *status quo* in the city or with promoting radical change. What characterises much, though by no means all, of the contemporary inner city debate is not, however, detailed argument as to the most appropriate policies to be followed within an agreed and accepted socio-political framework, but fundamental dispute as to the desirability of the frame-work itself.

Two related implications for the city follow from these observations. The first is that there can be no such thing as a 'correct' strategy for the inner city, rather, there are a set of alternatives, each of which will benefit some individuals and groups and disadvantage others. For example, existing policies favour established business and political interests in the city but do little in real terms to help the deprived, whose circumstances have deteriorated significantly in recent years. However, with a radical approach, low income groups may be expected to gain, but at the expense of the wealthy, whose privileged position will be undermined and diminished. In a society with a highly skewed distribution of wealth and income, any assessment of the overall effects of planning must, however, be conditioned by a recognition that 'profit' and 'loss' are relative and not absolute standards. What may be registered as a major gain for low income groups may represent only a minor loss to the affluent elite.

The second implication is that the various viewpoints on the inner city problem, though they can be researched in depth and discussed at length, cannot be resolved by academic argument. The reason is that they rest ultimately upon principles of ideological belief which are irreconcilable through scientific debate. Academic study can identify the problems of the city, and the most likely consequences of the courses of action which are envisaged for their solution, but choice of policy is invariably determined by political considerations and require-ments, rather than solely with respect to objective scientific criteria. Moreover, the arguments that may be expected to impress and influence policy makers the most are those which follow from the same ideo-logical premises which the policy makers themselves hold. For example, it is unlikely that a radical strategy will ever be supported, however powerful the scientific arguements in its favour, if those who manage the city are conservative in their political beliefs. Similarly, *laissez-faire* strategies are unlikely ever to appeal to those policy makers who are committed ideologically to promoting fundamental reform and change in society. The city is a product of choice, not chance. In voting for a particular political party, electorates are in fact choosing the type of city in which they wish to live.

Conclusion

Planning is state intervention in the city. It exists in order to alter and direct the city in ways decreed by society to be both necessary and desirable. Support for planning grew in the late nineteenth century in response to the problems of disease and overcrowding in the city. It led to legislation on building standards and layout and was implicit in a variety of utopian designs for neighbourhoods, communities and garden cities. As part of a broad-based social response, planning was seen in the early twentieth century as a means of achieving the new city and new society ideals which were expressed in the futurist, modern architecture, landscape design and City Beautiful movements. Despite common intellectual origins, practical planning as it existed in mid-century varied considerably from country to country. In the UK it consisted of a complex set of development controls and government action geared to the objectives of urban containment, protection of the countryside and new towns creation; whereas in the US, planning amounted to little more than zoning and the undertaking of public works. Today, it is insufficient to define planning simply in terms of land use and development regulation, rather it is part of a comprehensive urban policy which seeks to regulate a wide range of social and economic processes and relationships in the city. Changes over time and variations from place to place, however, preclude any general assessment of the effects of planning on the city. Equally, evaluation of its effects in any particular country is made difficult by the fact that it is not possible to say how the city would have developed without it. Containment, prevention of scatter, protection of the countryside and the success of the new towns number among the major achievements of post-war planning in the UK. Similarly, the separation of residential and industrial areas, and the public works programmes which have been undertaken are the most conspicuous products of zoning and planning in the USA. The recent emergence of serious inner city problems in both countries, however, is some indication that post-war planning has had its failings. Precisely how to arrest inner area decline remains the major issue facing contemporary urban planning in the Western world.

An appreciation of the effects of past planning policies on the city is clearly an important requirement in any branch of urban study. Indeed, in countries like the UK, which have highly developed planning systems, it is essential. In a reciprocal fashion, the most important practical value of urban study is that it provides an understanding of urban patterns and problems which is essential to the process of plan formation and policy selection. While they cannot 'solve' the problems of the city, urban geographers can contribute substantially and constructively to the urban debate. They can identify basic relationships, monitor ongoing processes, identify planning options and evaluate likely outcomes. Together with other specialists on the city they can help to ensure that the policies which are selected to shape the city of the future represent a deliberate, a considered and an informed choice.

The contemporary city poses exceptional challenges to the analyst, planner and policy maker. It requires a consideration of the complex mechanisms of growth, structure and management, topics which, because of their interrelatedness and complexity, transcend the limits and divisions of traditional academic inquiry. Given the need for an interdisciplinary approach it is clear that geographers have no monopoly of urban wisdom, indeed their interests and expertise are partial and restricted. What they contribute is a spatial perspective which is different from but which complements those insights provided by other urban specialists in the social and behavioural sciences. It has been the purpose of this guide to outline the ways in which the wider understanding of the city can be assisted and advanced by urban geographical study.

REFERENCES

Abercrombie, P. (1945) *The Greater London Plan*, London County Council, London

Abu-Lughod, J. (1961) 'Migrant adjustment to city life: the Egyptian case', *American Journal of Sociology, 67*, 22-32

Adams, R.M. (1960) 'The origin of cities', *Scientific American*, September

Alcaly, R.E. and Mermelstein, D. (1977) *The Financial Crisis of American Cities*, Vintage, New York

Alonso, W. (1964) *Location and Land Use*, Harvard University Press, Cambridge, Mass.

Ambrose, P. and Colenutt, B. (1975) *The Property Machine*, Penguin, Harmondsworth

Amedeo, D. and Golledge, R. (1975) *An Introduction to Scientific Reasoning in Geography*, Wiley, New York

Anderson, T.R. and Bean, L.L. (1961) 'The Shevky-Bell social areas: confirmation of results and a reinterpretation', *Social Forces, 40*, 119-24

Anderson, T.R. and Egeland, J. (1961) 'Spatial aspects of social area analysis', *American Sociological Review, 26*, 392-9

Appleyard, D. (1970) 'Styles and methods of structuring a city', *Environment and Behavior, 2*, 100-18

Armen, G. (1972) 'A classification of cities and city regions in England and Wales, 1966', *Regional Studies, 6*, 149-82

Aurousseau, M. (1924) 'Recent contributions to urban geography', *Geographical Review, 14*, 444-55

Badcock, B.A. (1970) 'Central place evolution and network development in South Auckland 1840-1968', *The New Zealand Geographer, 26*, 109-35

Bailey, F.G. (1957) *Caste and the Economic Frontier*, University Press, Manchester

Bailly, A.S. and Polèse, A. (1977) 'Processus urbains et modèles spatiaux: ecologie factorielle comparée Edmonton-Quebec', *Canadian Geographer, 21*, 59-80

Barnum, H.G. (1966) *Market Centres and Hinterlands in Baden-Württemberg*, Department of Geography Research Paper 103, University of Chicago, Chicago

Bassett, K. and Hauser, D. (1976) 'Public policy and spatial structure: housing improvement in Bristol' in R.F. Peel, M.D.I. Chisholm and P. Haggett (eds.), *Processes in Physical and Human Geography: Bristol Essays*, Heinemann, London

Bassett, K. and Short J.R. (1978) 'Housing improvement in the inner city: a case study of changes before and after the 1974 Housing Act', *Urban Studies, 15*, 333-42

Bassett, K. and Short, J.R. (1980) *Housing and Residential Structure*, Routledge & Kegan Paul, London

Batty, M. (1976) *Urban Modelling: Algorithms, Calibrations, Predictions*, Cambridge University Press, London

Bell, W. (1955) 'Reply to Duncan's comments on social area analysis', *American Journal of Sociology, 61*, 260-2

—— (1965) 'Economic, family and ethnic status: an empirical test', *American*

Sociological Review, 20, 45-52

Berger, B. (1968) 'Myths of American suburbia' in R.E. Pahl (ed.), *Readings in Urban Sociology,* Pergamon, London, pp. 119-35

Berry, B.J.L. (1961) 'City size distributions and economic development', *Economic Development and Cultural Change, 9,* 573-88

————— (1964) 'Approaches to regional analysis: a synthesis', *Annals, Association of American Geographers, 54,* 2-11

————— (1966) *Essays on Commodity Flows and the Spatial Structure of the Indian Economy,* Department of Geography Research Paper 111, University of Chicago, Chicago

————— (1967) *Geography of Market Centres and Retail Distribution,* Prentice Hall, Englewood Cliffs, NJ

————— (1970) 'The U.S.A. in the year 2000', *Transactions Institute of British Geographers, 51,* 21-53

————— (1971) 'Comparative factorial ecology', *Economic Geography,* (supplement), 47

————— (1972) *City Classification Handbook: Methods and Applications,* Wiley, New York

————— (1973) *The Human Consequences of Urbanisation,* St Martin's Press, New York

Berry, B.J.L. and Garrison, W.L. (1958a) 'The functional bases of the central place hierarchy', *Economic Geography, 34,* 145-54

————— (1958b) 'Recent developments in central place theory', *Papers & Proceedings, Regional Science Association, 4,* 107-20

Berry, B.J.L. and Horton, F.E. (1969) 'Emergence of urban geography: methodological foundations of the discipline' in B.J.L. Berry and F.E. Horton (eds.), *Geographic Perspectives on Urban Systems,* Prentice Hall, Englewood Cliffs, NJ, pp. 1-19

Berry, B.J.L. and Rees, P.H. (1969) 'The factorial ecology of Calcutta', *American Journal of Sociology, 74,* 445-91

Bird, H. (1976) 'Residential mobility and preference patterns in the public sector of the housing market', *Transactions, Institute of British Geographers, NS 1,* 20-33

Blake, P. (1977) 'Britain's new towns: facts and figures', *Town and Country Planning, 45,* 89-103

Blanchard, R. (1911) 'Grenoble', *Etude de Geographie Urbaine,* Paris

Blaut, J.M., McCleary, G.F. and Blaut, A.S. (1970) 'Environmental mapping in young children', *Environment and Behavior, 2,* 335-49

Boddy, M. (1976) 'The structure of mortgage finance: building societies and the British social formation', *Transactions, Institute of British Geographers, NS 1,* 20-33

————— (1980) 'Finance capital commodity production and the production of urban built form' in M.J. Dear and A.J. Scott (eds.), *Urbanisation and Urban Planning in Capitalist Societies,* Maaroufa Press, Chicago

Booth, C. (1889-1903) *Life and Labour of the People of London,* Macmillan, London

Bopegamage, A. (1957) 'Neighbourhood relations in Indian cities – Delhi', *Sociological Bulletin, 6,* 34-42

Borchert, J.R. (1967) 'American metropolitan evolution', *Geographical Review,*

57, 301-23

——— (1972) 'America's changing metropolitan regions', *Annals, Association of American Geographers, 62,* 355-6

Bourne, L. and Barber, G. (1971) 'Ecological patterns of small urban centres in Canada' in B.J.L. Berry (ed.), 'Comparative factorial ecology', *Economic Geography, 47,* (supplement), 258-65

Bowman, R.A. and Hoshing, P.L. (1971) 'A factorial ecology of Auckland' in R.J. Johnston and J.M. Soons (eds.), *Proceedings of the Sixth New Zealand Geography Conference,* vol. 1, New Zealand Geographical Society, Christchurch, pp. 273-80

Briggs, A. (1963) *Victorian Cities,* Odhams, London

Brush, J.E. (1953) 'The hierarchy of central places in Southwestern Wisconsin', *Geographical Review, 43,* 380-402

Buckingham, J.S. (1849) *National Evils and Practical Remedies,* Cadell and Davies, London

Bull, P.J. (1978) 'The spatial components of intra-urban manufacturing change: suburbanisation in Clydeside 1958-1968', *Transactions of the Institute of British Geographers, NS 3,* 91-100

Bunge, W. (1975) 'Detroit humanly viewed: the American urban present' in R. Abler, D.G. Janelle, A.K. Philbrick and J. Sommers (eds.), *Human Geography in a Shrinking World,* Duxbury, North Scituate, Mass.

Burgess, E.W. (1925) 'The growth of the city' in R.E. Park, E.W. Burgess and R.D. McKenzie (eds.), *The City,* University of Chicago Press, Chicago, pp. 47-62. Reprinted in Stewart, M. (1972) *The City: Problems of Planning,* Penguin, Harmondsworth

Canter, D.V. and Tagg, S.K. (1975) 'Distance estimation in cities', *Environment and Behavior, 7,* 59-80

Carey, G.W. (1966) 'The regional interpretation of population and housing patterns in Manhattan through factor analysis', *Geographical Review, 46,* 551-69

Carey, G.W., Macomber, L. and Greenberg, M. (1968) 'Educational and demographic factors in the urban geography of Washington, D.C.', *Geographical Review, 48,* 515-37

Carter, H. (1972) *The Study of Urban Geography,* Edward Arnold, London

Catanese, A.J. (1979) 'History and trends of urban planning' in A.J. Catanese and J.C. Snyder, *Introduction to Urban Planning,* McGraw-Hill, London

CDP (1975) *Coventry and Hillfields: Prosperity and the Persistence of Inequality (Final Report),* Community Development Project, Coventry

Chabot, G. (1938) 'La détermination des courbes isochrones én geographie urbaine', *Comptes Rendus du Congrès Internationale de Géographie, tome 2: Géographie Humaine,* 110-13

Chandler, A.D. and Redlich, F. (1961) 'Recent developments in American business administration and their conceptualisation', *Business History Review,* Spring, 103-28

Cherry, G.E. (1974) *The Evolution of British Town Planning,* Leonard Hill, London

Childe, V.G. (1950) 'The urban revolution', *Town Planning Review, 21,* 3-17

Chorley, P.J. and Haggett, P. (1967) *Models in Geography,* Methuen, London

Christaller, W. (1933) *Die Zentralen Orte in Suddeutschland,* Gustav Fischer,

Jena. translated by Baskin, C.W. (1966) as *Central Places in Southern Germany*, Prentice-Hall, Englewood Cliffs, NJ

Clark, C. (1940) *The Conditions of Economic Progress*, Macmillan, London

―――― (1951) 'Urban population densities', *Journal of the Royal Statistical society, Series A, 114*, 490-6

Clark, D. (1973) 'Urban linkage and regional structure in Wales: An analysis of change, 1958-68' *Transactions, Institute of British Geographers, 58,* 41-58

―――― (1979) 'The spatial impact of telecommunication' in R.C. Smith (ed.), *Impacts of Telecommunications on Planning and Transport*, Departments of the Environment and Transport Research Report 24, London, pp. 85-128

Clark, D., Davies, W.K.D. and Johnston, R.J. (1974) 'The application of factor analysis in human geography', *The Statistician, 23,* 259-81

Clark, P.J. and Evans, F.C. (1954) 'Distance to nearest neighbour as a measure of spatial relationships in populations', *Ecology, 35,* 445-53

Conzen, M.R.G. (1960) *Alnwick, Northumberland: A Study in Town Plan Analysis*, Institute of British Geographers Monograph, London

―――― (1962) 'The plan analysis of an English city centre' in K. Norborg (ed.), *Proceedings of the I.G.U. Symposium in Urban Geography*, C.W.K. Gleerup, Lund

Cox, K.R. (1973) *Conflict, Power and Politics in the City: A Geographic View*, McGraw Hill, New York

Cox, K.R. and Golledge, R.G. (1969) 'Editorial Introduction: behavioral models in geography' in K.R. Cox and R.G. Golledge (eds.), *Behavioral Problems in Geography: A Symposium*, Studies in Geography 17, Northwestern University, Evanston, pp. 1-13

Curry, L. (1964) 'The random spatial economy: an exploration in settlement theory', *Annals, Association of American Geographers, 54,* 138-46

―――― (1967) 'Central places in the random spatial economy', *Journal of Regional Science, 7* (supplement), 217-38

Dacey, M.F. (1966) 'A probability model for central place locations', *Annals, Association of American Geographers, 56,* 549-68

Dale, P.F. (1971) 'Children's reaction to mental maps and aerial photographs', *Area, 3,* 170-7

Daniels, P.W. (1975) *Office Location: An Urban and Regional Study*, Bell, London

―――― (1979) *Spatial Patterns of Office Growth and Location*, Wiley, New York

Darwin, C. (1859) *The Origin of Species*, The New English Library, London (1958)

Davies, W.K.D. (1967) 'Centrality and the central place hierarchy', *Urban Studies, 4,* 61-79

―――― (1968) 'Morphology of central places: a case study', *Annals, Association of American Geographers, 58,* 91-110

―――― (1970) 'Towards an integrated study of central places: a South Wales case study' in H. Carter and W.K.D. Davies (eds.), *Urban Essays: Studies in the Geography of Wales*, Longman, London, pp. 193-227

―――― (1972) 'Geography and the methods of modern science' in W.K.D. Davies (ed.), *The Conceptual Revolution in Geography*, University of London Press, London, pp. 131-9

―――― (1975) 'Variance allocation and the dimensions of British towns', *Tijdschrift voor Economische en Sociale Geografie, 66,* 358-72

Davies, W.K.D. and Barrow, G.T. (1973) 'A comparative factorial ecology of three Prairie cities', *Canadian Geographer, 17*, 327-53

Davies, W.K.D. and Lewis, G.J. (1973) 'The urban dimensions of Leicester' in B. Clarke and B. Gleave (eds.), *Social Patterns in Cities*, Special Publications 5, Institute of British Geographers, London, pp. 71-86

Davies, W.K.D. and Thompson, R.R. (1980) 'The structure of interurban connectivity: a dyadic factor analysis of Prairie commodity flows', *Regional Studies, 14*, 297-311

Davis, K. (1965) 'The urbanisation of the human population', *Scientific American, 213*, 40-53

―――― (1973) 'World urbanisation, 1950-70' in *World Urbanisation*, University of California Population Monograph Series No. 9, Berkeley. Reprinted in L.S. Bourne and J.W. Simmons (eds.) (1978) *Systems of Cities*, Oxford University Press, New York, pp. 92-9

Delafons, J. (1969) *Land Use Controls in the United States*, MIT Press, Cambridge, Mass.

Dennis, N. (1978) 'Housing policy areas: criteria and indicators in principle and practice', *Transactions, Institute of British Geographers, NS 3*, 2-22

Department of the Environment (1977a) *Change or Decay? Final Report of the Liverpool Inner Area Study*, HMSO, London

―――― (1977b) *Inner Area Studies: Summaries of Consultants' Final Reports*, HMSO, London

Dickinson, R.E. (1947) *City, Region and Regionalism*, Routledge & Kegan Paul, London

―――― (1948) 'The scope and status of urban geography: an assessment', *Land Economics, 24*, 221-38. Reprinted in H.M. Mayer and C.F. Kohn (eds.) (1959) *Readings in Urban Geography*, University of Chicago Press, Chicago, pp. 10-26

Dijkink, G. and Elbers, E. (1981) 'The development of geographic representation in children', *Tijdschrift voor Economische en Sociale Geografie, 72*, 2-16

Downs, R.M. and Stea, D. (1973) *Image and Environment*, Edward Arnold, London

―――― (1977) *Maps in Mind*, Harper and Row, New York

Doxiadis, C.A. (1966) *Urban Renewal and the Future of the American City*, Public Administration Service, Chicago

Drewett, R., Goddard, J.B., Kennett, S. and Spence, N.A. (1975) 'The standard metropolitan labour area and metropolitan economic labour area concepts applied to Northern Ireland: Population analysis and comparison with Great Britain', Department of Geography Working Paper No. 19, London School of Economics and Political Science, London

Drewett, R., Goddard, J.B., Spence, N.A., Pinch, S., Thornton, S. and Williams, A. (1976) *Socio-economic Change in the British Urban System, 1961-71*, Department of Geography Working Paper No. 21, London School of Economics and Political Science, London

Dubuc, R. (1938) 'L'Apportionnement de Paris en lait', *Annales de Geographie, 47*, 257-66

Duncan, S.S. (1974) 'Cosmetic planning or social engineering?', *Area, 6*, 259-70

―――― (1976) 'Self help: the allocation of mortgages and the formation of housing sub-markets', *Area, 8*, 307-16

Durkheim, E. (1893) *De la Division du Travail Social*. Translated by G. Simpson, as *The Division of Labour in Society*, Free Press, New York

Evans, A.W. (1973) *The Economics of Residential Location*, Macmillan, London

Evans, D.J. (1973) 'Urban social structures in South Wales' in B. Clarke and B. Gleave (eds.), *Social Patterns in Cities*, Special Publication 5, Institute of British Geographers, London, pp. 87-102

Everitt, J. and Cadwallader, M. (1972) 'The home area concept in urban analysis' in W.J. Mitchell (ed.), *Environmental Design: Research and Practice*, University of California, Los Angeles

Everson, J.A. and Fitzgerald, B.P. (1969) *Settlement Patterns*, Longman, London

Fawcett, C.B. (1932) 'Distribution of the urban population in Great Britain', *Geographical Journal, 79*, 97-106

Fischer, C.S. (1976) *The Urban Experience*, Harcourt Brace Jovanovich, New York

Ford, J. (1975) 'The role of the building society manager in the urban stratification system: autonomy versus constraint', *Urban Studies, 12*, 295-302

Francescato, D. and Mebane, W. (1973) 'How citizens view two great cities: Milan and Rome' in R. Downs and D. Stea (eds.), *Images and Environment*, Aldine, Chicago, pp. 131-47

Gans, H.J. (1962a) 'Urbanism and suburbanism as ways of life' in A.M. Rose (ed.), *Human Behavior and Social Processes*, Routledge & Kegan Paul, London. Reprinted in R.E. Pahl (ed.) (1968), *Readings in Urban Sociology*, Pergamon, London

—— (1962b) *The Urban Villagers*, Free Press, New York

—— (1967) *The Levittowners*, Pantheon, New York

Geddes, P. (1915) *Cities in Evolution*, Williams and Norgate, London

Geisler, W. (1924) 'Die Deutsche Stadt', *Forschungen zu Deutschen Lande und Volkskunde*, Stuttgart

Gibbs, J.P. and Schnore, L.E. (1960) 'Metropolitan growth: an international study', *American Journal of Sociology, 66*, 160-78

Giggs, J.A. (1970) 'Socially disorganised areas in Barry: a multivariate analysis' in H. Carter and W.K.D. Davies (eds.), *Urban Essays: Studies in the Geography of Wales*, Longman, London, pp. 101-43

Goddard, J.B. (1973) *Office Linkages and Location*, Progress in Planning 1, Pergamon, Oxford

—— (1975) *Office Location in Urban and Regional Development*, Pergamon, Oxford

Goddard, J.B. and Smith, I.J. (1978) 'Changes in corporate control in the British urban system, 1972-7', *Environment and Planning A, 10*, 1073-84

Gold, J.R. (1980) *An Introduction to Behavioural Geography*, Oxford University Press, Oxford

Golledge, R.G., Rushton, G. and Clark, W.A.V. (1966) 'Some spatial characteristics of Iowa's dispersed farm population and their implications for the grouping of central place functions', *Economic Geography, 43*, 261-72

Goodchild, B. (1974) 'Class differences in environmental perception', *Urban Studies, 11*, 157-69

Goodey, B. and Lee, S.A. (1971) *The City Scope Project, Hull 1971*, Report for McKinsey & Co., London

Goodey, B., Duffett, A.W., Gold, J.R. and Spender, D. (1971) 'City scene: an exploration into the image of central Birmingham', Centre for Urban and

220   *References*

Regional Studies Research Memorandum No. 10, University of Birmingham, Birmingham

Gottmann, J. (1961) *Megalopolis: The Urbanised Northeastern Seaboard of the United States*, The Twentieth Century Fund, New York

—— (1976) 'Megalopolitan systems around the world', *Ekistics, 243*, 109-13. Reprinted in L.S. Bourne and J.W. Simmons (eds.) (1978) *Systems of Cities*, Oxford University Press, New York, pp. 53-60

Gould, P.R. (1972) 'Pedagogic review', *Annals, Association of American Geographers, 62*, 689-700

Gould, P.R. and White, R. (1974) *Mental Maps*, Penguin Books, Harmondsworth

Gray, F. (1976) 'Selection and allocation in council housing', *Transactions, Institute of British Geographers, NS 1*, 34-46

Greenhut, M.L. (1956) *Plant Location in Theory and Practice: The Economics of Space*, University of North Carolina Press, Chapel Hill

Grigsby, W. (1970) 'Home finance and housing quality in aging neighbourhoods' in M. Stegman (ed.), *Housing and Economics: The American Dilemma*, MIT Press, Cambridge, Mass., pp. 299-316

Gripaios, P. (1977) 'The closure of firms in the inner city: the South-East London case, 1970-5', *Regional Studies, 11*, 1-6

Haggett, P. (1975) *Geography: A Modern Synthesis*, Harper and Row, London

Haig, R.M. (1926) 'Towards an understanding of the metropolis', *Quarterly Journal of Economics, 1*, 179-208 and *3*, 402-34

Hall, P. (1966) *The World Cities*, McGraw Hill, New York

Hall, P., Gracey, H., Drewett, R. and Thomas, R. (1973) *The Containment of Urban England*, George Allen and Unwin, London

Hamnett, C. (1979) 'Area-based explanations: a critical appraisal' in D.T. Herbert and D.M. Smith (eds.), *Social Problems and the City: Geographical Perspectives*, Oxford University Press, Oxford

Hannah, L. (1976) *The Rise of the Corporate Economy*, Methuen, London

Harloe, M., Issacharoff, R. and Minns, R. (1974) *The Organisation of Housing*, Heinemann, London

Harrison, J. and Howard, W. (1972) 'The role of meaning in the urban image', *Environment and Behaviour, 4*, 389-411

Hartshorne, R. (1939) *The Nature of Geography*, Association of American Geographers, Lancaster, Pa

Harvey, D.W. (1969) *Explanation in Geography*, Edward Arnold, London, p. 34

—— (1973) *Social Justice and the City*, Edward Arnold, London

Hassert, K. (1907) *Die Stadte Geographisch Betrachtet*, Veit, Leipzig

Hawley, A.H. and Duncan, O.D. (1957) 'Social area analysis: a critical appraisal', *Land Economics, 33*, 337-45

Herbert, D.T. (1967) 'Social area analysis: a British study', *Urban Studies, 4*, 41-60

—— (1970) 'Principal components analysis and urban social structure: a study of Cardiff and Swansea' in H. Carter and W.K.D. Davies (eds.), *Urban Essays: Studies in the Geography of Wales*, Longman, London, pp. 79-100

—— (1972) *Urban Geography: A Social Perspective*, David and Charles, London

Herbert, D.T. and Johnston, R.J. (1978) 'Geography and the urban environment' in D.T. Herbert and R.J. Johnston (eds.), *Geography and the Urban Environment: Progress in Research and Applications, Volume 1*, Wiley, London, pp. 1-29

Hill, R.C. (1977) 'Fiscal collapse and political struggle in decaying central cities in the United States' in W.K. Tabb and L. Sawers (eds.), *Marxism and the Metropolis, New Perspectives in Urban Political Economy,* Oxford University Press, New York, pp. 213-40

HMSO (1977) *Policy for the Inner Cities,* Cmnd. 6845, HMSO, London

Hoover, E.M. (1948) *The Location of Economic Activity,* McGraw-Hill, New York

Horst, P. (1965) *Factor Analysis of Data Matrices,* Holt, Rinehart and Winston, New York

Howard, E. (1898 and 1902) *Tomorrow: A Peaceful Path to Real Reform* (1898), republished as *Garden Cities of Tomorrow* (1902). Reprinted by Faber and Faber (F.J. Osborn (ed.)), 1946

Hoyt, H. (1939) *The Structure and Growth of Residential Neighbourhoods in American Cities,* Federal Housing Administration, Washington, DC

Hurd, R.M. (1903) *Principles of City Land Values,* The Record and Guide, New York

Hymer, S. (1972) 'The multinational corporation and the law of uneven development' in J.N. Bhagwati (ed.), *Economics and World Order,* Macmillan, London

Isard, W. (1956) *Location and Space-economy: a General Theory Relating to Industrial Location, Market Areas, Land Use, Trade and Urban Structure,* John Wiley, New York

James, P.E. (1972) *All Possible Worlds: A History of Geographical Ideas,* The Odyssey Press, Indianapolis

Janson, C. (1968) 'The spatial structure of Newark, New Jersey. Part I: The Central City', *Acta Sociologica, 11,* 144-69

Jefferson, M. (1939) 'The law of the primate city', *Geographical Review, 29,* 226-32

Johnston, R.J. (1964) 'The measurement of a hierarchy of central places', *Australian Geographer, 9,* 315-17

—— (1973a) 'Neighbourhood patterns within urban areas' in R.J. Johnston *Urbanisation in New Zealand,* Reed, Auckland, pp. 204-27

—— (1973b) 'Residential differentiation in major New Zealand urban areas: a comparative factorial ecology' in B. Clarke and B. Gleave (eds.), *Social Patterns in Cities,* Special Publication 5, Institute of British Geographers, London, pp. 143-67

—— (1976) 'Residential area characteristics: research methods for identifying urban sub-areas – social area analysis and factorial ecology' in D.T. Herbert and R.J. Johnston (eds.), *Social Areas in Cities,* vol. 1, John Wiley, London

Jones, F.L. (1969) *Dimensions of urban social structure: the social areas of Melbourne, Australia,* Australian National University Press, Canberra

Jones, P.N. (1979) 'Ethnic areas in British cities' in D.T. Herbert and D.M. Smith (eds.), *Social Problems and the City,* Oxford University Press, Oxford

Jonge, D. De (1962) 'Images of urban areas: their structure and psychological foundations', *Journal, American Institute of Planners, 28,* 266-76

Kahn, H. and Weiner, A.J. (1967) *The Year 2000,* Macmillan, New York

Kain, J.F. (1962) 'The journey to work as a determinant of residential location', *Papers and Proceedings, Regional Science Association, 9,* 137-60

Karn, V. (1976) 'Priorities for local authority mortgage lending: a case study of Birmingham', Centre for Urban and Regional Studies Research Memorandum No. 52, University of Birmingham, Birmingham

King, L.J. (1962) 'A quantitative expression of the pattern of urban settlements in selected areas of the United States', *Tijschrift voor Economische en Sociale Geografie, 53,* 1-7. Reprinted in B.J.L. Berry and D.F. Marble (1968), *Spatial Analysis,* Prentice Hall, Englewood Cliffs, NJ, pp. 159-67

King, L.J. and Golledge, R.G. (1978) *Cities, Space and Behavior: The Elements of Urban Geography,* Prentice Hall, Englewood Cliffs, NJ

Kirby, A. (1978) *The Inner City: Causes and Effects,* Retail and Planning Associates, Newcastle-upon-Tyne

Labasse, J. (1955) *Les Capitaux et la Region: Etude geographique essai sur le commerce et la circulation des capitaux dans la region Lyonnaise,* Cahiers de la Fondation Nationale des Sciences Politiques, Paris, p. 69

Ladd, F.C. (1970) 'Black youths view their environment: neighbourhood maps', *Environment and Behavior, 2,* 74-99

Lambert, C. (1976) 'Building societies, surveyors, and the older areas of Birmingham', Centre for Urban and Regional Studies Working Paper No. 38, University of Birmingham, Birmingham

Lambert, J. (1970) 'The management of minorities', *New Atlantis, 2,* 49-79

Lampard, E.E. (1965) 'Historical aspects of urbanisation' in P.M. Hauser and L.F. Schnore (eds.), *The Study of Urbanisation,* John Wiley, London, pp. 519-54

Lawless, P. (1981) *Britain's Inner Cities: Problems and Policies,* Harper and Row, London

Le Corbusier (1925) *Towards a New Architecture,* Architectural Press, London

Lewis, C.R. (1970) 'The central place pattern of Mid-Wales and the middle Welsh borderland' in H. Carter and W.K.D. Davies (eds.), *Urban Essays: Studies in the Geography of Wales,* Longman, London, pp. 228-68

Lewis, O. (1952) 'Urbanisation without breakdown', *Scientific Monthly, 75,* 31-41

Littlejohn, I. (1963) *Westrigg: The Sociology of a Cheviot Parish,* Routledge & Kegan Paul, London

Lloyd, P.E. and Mason, C.M. (1978) 'Manufacturing industry in the inner city: a case study of Greater Manchester', *Transactions of the Institute of British Geographers, NS 3,* 66-90

Lösch, A. (1954) *The Economics of Location,* Yale University Press, New Haven, Conn.

Luckermann, F. (1966) 'Empirical expressions of nodality and hierarchy in a circulation manifold', *East Lakes Geographer, 2,* 17-44

Lynch, K. (1960) *The Image of the City,* MIT Press, Cambridge, Mass.

Manchester Area Rapid Transit Study (1967) *Volume 1: Report of the Working Party,* Town Hall, Manchester

Marshall, J.U. (1969) *The Location of Service Towns,* Department of Geography Research Publication No. 3, University of Toronto, Toronto

Martiny, R. (1928) 'Grundrissgestaltung der deutschen Siedlungen', *Petermanns Mitteilungen, Ergänszungschaft,* 197

Martyn, W.F. (1793) *The Geographical Magazine: or New System of Geography,* John Stockdale, London

Matthews, M.H. (1980) 'The mental maps of children: images of Coventry's city centre', *Geography, 65,* 169-79

Maurer, R. and Baxter, J.C. (1972) 'Images of the neighbourhood and city among black-, Anglo-, and Mexican-American children', *Environment and Behavior, 4,* 351-88

Mayer, H.M. (1965) 'A survey of urban geography' in P.M. Hauser and L.F. Schnore (eds.), *The Study of Urbanisation,* Wiley, London

Mayer, P. (1962) 'Migrancy and the study of Africans in towns', *American Anthropologist, 64,* 576-91

—— (1963) *Townsmen or Tribesmen,* Oxford University Press, Cape Town

Mayfield, R. (1967) 'A central place hierarchy in Northern India' in W.L. Garrison and D.F. Marble (eds.), *Quantitative Geography, Part 1: Economic and Cultural Topics,* Department of Geography Study No. 13, Northwestern University, Evanston, Illinois

Mayhew, H. (1861) *London Labour and the London Poor,* 4 vols, Constable, London

Maxwell, J.W. (1965) 'The functional structure of Canadian cities: a classification of cities', *Geographical Bulletin, 7,* 79-104

McElrath, D. (1962) 'The social areas of Rome: a comparative analysis', *American Sociological Review, 27,* 376-91

McGee, T.G. (1969) 'The social ecology of New Zealand cities' in J. Forster (ed.), *Social Processes in New Zealand,* Longman Paul, Auckland

McKenzie, R.D. (1925) 'The ecological approach to the study of the human community' in R.E. Park, E.W. Burgess and R.D. McKenzie (eds.), *The City,* University of Chicago Press, Chicago, pp. 63-79

McLuhan, H.M. (1962) *The Gutenberg Galaxy: The Making of Typographic Man,* Routledge & Kegan Paul, London

—— (1964) *Understanding Media,* McGraw-Hill, New York

Meier, R.L. (1962) *A Communications Theory of Urban Growth,* MIT Press, Cambridge, Mass.

Michaelson, W. (1970) *Man and his Enviroment: A Sociological Approach,* Addison and Wesley, London

Milgram, S., Greenwald, J., Kessler, S., McKenna, W. and Waters, J. (1972) 'A psychological map of New York City', *American Scientist, 60,* 194-200

Mills, E.S. (1972) *Studies in the Structure of the Urban Economy,* John Hopkins Press, Baltimore, Md.

Müller, E. (1931) *Die Altstadt von Breslau: City Bildung und Physiognomie,* Veröff, der Schlesischen Ges. für Erdkunde, Breslau

Muller, P.O. (1981) *Contemporary Suburban America,* Prentice Hall, Englewood Cliffs, NJ

Murdie, R.A. (1969) *Factorial Ecology of Metropolitan Toronto, 1951-61,* Department of Geography Research Paper 116, University of Chicago, Chicago

—— (1971) 'The social geography of the city: theoretical and empirical background' in L.S. Bourne (ed.), *The Internal Structure of the City,* Oxford University Press, Oxford, pp. 279-90

Muth, R.F. (1969) *Cities and Housing,* Chicago University Press, Chicago

National Resources Planning Board (1937) *Our Cities: Their Role in the National Economy,* National Resources Planning Board, Washington, DC

—— (1939) *Urban Planning and Land Policies,* National Resources Planning Board, Washington, DC

Openshaw, S. (1975) *Some Theoretical and Applied Aspects of Spatial Interaction Shopping Models,* Geo-Abstracts Ltd., Norwich

Orleans, P. and Schmidt, S. (1972) 'Mapping the city: environmental cognition of urban residents', *Environmental Design Research Association, 3,* 1-9

Owen, R. (1813) *A New View of Society,* Cadell and Davies, London

Pahl, R.E. (1965) *Urbs in Rure: The Metropolitan Fringe in Hertfordshire,* Geographical Papers No. 2, London School of Economics and Political Science, London

—— (1968) *Readings in Urban Sociology,* Pergamon, Oxford

—— (1970) *Patterns of Urban Life,* Longman, London

—— (1975) *Whose City?* Penguin, Harmondsworth

—— (1979) 'Socio-political factors in resource allocation' in D.T. Herbert and D.M. Smith (eds.), *Social Problems and the City,* Oxford University Press, Oxford, pp. 33-41

Palm, R.I. (1976a) 'The role of real estate agents as information mediators in two American cities', *Geografiska Annaler, 58B,* 28-41

—— (1976b) 'Real estate agents and geographical information', *Geographical Review, 66,* 266-80

—— (1979) 'Financial and real estate institutions in the housing market: a study of recent house price changes in the San Francisco Bay area' in D.T. Herbert and R.J. Johnston (eds.), *Geography and the Urban Environment, Progress in Research and Applications,* vol. II, John Wiley, London, pp. 83-124

Paris, C. (1974) 'Urban renewal in Birmingham, England: an institutional approach', *Antipode, 6,* 7-15

Paris, C. and Lambert, J. (1979) 'Housing problems and the state: the case of Birmingham, England' in D.T. Herbert and R.J. Johnston (eds.), *Geography and the Urban Environment. Progress in Research and Applications,* vol. II, John Wiley, London, pp. 227-58

Park, R.E. (1916) 'The city: suggestions for the investigation of human behavior in the urban environment', *American Journal of Sociology, 20,* 577-612. Reprinted in R.E. Park, E.W. Burgess and R.D. McKenzie (eds.), *The City,* University of Chicago Press, Chicago, pp. 1-46

—— (1936) 'Human ecology', *American Journal of Sociology, 62,* 1-15. Reprinted in G.A. Theodorson (ed.) (1961) *Studies in Human Ecology,* Harper and Row, New York

Parkes, D.N. (1972) 'A classical social area analysis: Newcastle and some comparisons', *The Australian Geographer, 11,* 555-78

Peach, G.C.K. (1975) 'Immigrants in the inner city', *Geographical Journal, 141,* 372-9

Peet, R. (1978) *Radical Geography,* Methuen, London

Pinkerton, J. (1807) *Modern Geography: A Description of the Empires, Kingdoms, States and Colonies: With the Oceans, Seas and Islands: In All Parts of the World,* Cadell and Davies, London

Pocock, D.C.D. (1975) 'Durham: images of a cathedral city', Department of Geography Occasional Publications No. 6, University of Durham, Durham

Pocock, D.C.D. and Hudson, R. (1978) *Images of the Urban Environment,* Macmillan, London

Porteous, J.D. (1971) 'Design with people: the quality of the urban environment', *Environment and Behavior, 3,* 155-78

—— (1977) *Environment and Behavior: Planning and Everyday Urban Life,*

Addison and Wesley, London

Pred, A.R. (1966) *The Spatial Dynamics of U.S. Urban – Industrial Growth 1800-1914*, MIT Press, Cambridge, Mass.

—— (1967) *Behavior and Location: Foundations for a Geographic and Dynamic Location Theory*, Part I, C.W.K. Gleerup, Lund

—— (1969) *Behavior and Location: Foundations for a Geographic and Dynamic Location Theory*, Part II, C.W.K. Gleerup, Lund

—— (1977) *City Systems in Advanced Economies*, Hutchinson, London

Preston, R.E. (1971) 'The structure of central place systems', *Economic Geography, 47*, 145-7

—— (1979) 'The recent evolution of Ontario central place systems in the light of Christaller's concept of centrality', *Canadian Geographer, 23*, 201-21

Ratcliffe, J. (1974) *An introduction to Town and Country Planning*, Hutchinson, London

Reader, W.J. (1975) *Imperial Chemical Industries: A History*, Weidenfeld & Nicolson, London

Redfield, R. (1941) *Folk Culture of Yucatan*, University of Chicago Press, Chicago

Rees, A.D. (1950) *Life in a Welsh Countryside*, University of Wales Press, Cardiff

Rees, P.H. (1970) 'The factorial ecology of metropolitan Chicago, 1960' in B.J.L. Berry and F.L. Horton (eds.), *Geographic Perspectives on Urban Systems*, Prentice Hall, Englewood Cliffs, NJ

Rees, P.H. and Wilson, A.G. (1977) *Spatial Population Analysis*, Arnold, London

Reissman, L. (1964) *The Urban Process*, Free Press, New York

Rhind, D.W., Evans, I.S. and Visvalingam, M. (1980) 'Making a national atlas of population by computer', *The Cartographic Journal, 17*, 3-11

Richardson, H.W. (1977) *The New Urban Economics*, Pion, London

Robson, B.T. (1969) *Urban Analysis*, Cambridge University Press, Cambridge

Romanos, M.C. (1976) *Residential Spatial Structure*, Heath, New York

Rowley, G. (1970) 'Central places in rural Wales: a case study', *Tijschrift voor Economische en Sociale Geografie, 61*, 32-41

—— (1971) 'Central places in rural Wales', *Annals, Association of American Geographers, 61*, 537-50

Rummell, R.J. (1967) 'Understanding factor analysis', *Journal of Conflict Resolution, 11*, 444-80

Rushton, G. (1969) 'Analysis of spatial behavior by revealed space preference', *Annals, Association of American Geographers, 59*, 391-400

Saarinen, T.F. (1969) *Perception of environment*, Commission on College Geography Resource Paper No.5, Association of American Geographers, Washington

Samuels, J.M. (1965) 'Size and growth of firms', *Economic Studies Review, 32*, 105-12

Schmid, C.F. and Tagashira, K. (1964) 'Ecological and demographic indices: a methodological analysis', *Demography, 1*, 194-211

Scott, P. (1964) 'The hierarchy of central places in Tasmania', *Australian Geographer, 9*, 134-7

Shannon, H.A. (1931) 'The coming of general limited liability', *Economic History, 2*, 31-46

Shevky, E. and Bell, W. (1955) *Social Area Analysis*, Stanford University Press, Stanford

Shevky, E. and Williams, M. (1949) *The Social Areas of Los Angeles*, University of California Press, Los Angeles

Short, J. (1977) 'Social system and spatial patterns', *Antipode, 8,* 77-87

Simmons, J.W. and Bourne, L.S. (1978) 'Defining urban places: differing concepts of the urban system' in L.S. Bourne and J.W. Simmons (eds.), *Systems of Cities: Readings on Structure, Growth and Policy,* Oxford University Press, New York

Simnel, G. (1903) *Die grosstadt und das geistesleben.* Translated and reprinted as 'The Metropolis and urban life' in P.K. Hatt and A.J. Reiss (eds.) (1951), *Reader in Urban Sociology,* Free Press, New York

Simon, H.A. (1955) 'On a class of skew distribution functions', *Biometrica, 42,* 425-40

Sjoberg, G. (1960) *The Pre-industrial City,* Free Press, New York

Stacey, M. (1960) *Tradition and Change,* Oxford University Press, Oxford

Stamp, L.D. (1964) *The Geography of Life and Death,* Collins, London

Suttles, G.D. (1975) 'Community design: the search for participation in a metropolitan society' in A.H. Hawley and V.P. Rock (eds.), *Metropolitan America in Contemporary Perspective,* Halstead, New York, pp. 235-97

Sweetzer, F.L. (1965) 'Factor structure as ecological structure in Helsinki and Boston', *Acta Sociologica, 8,* 205-25

Taaffe, E.J. (1956) 'Air transportation and the United States urban distribution', *Geographical Review, 46,* 219-38

────── (1962) 'The urban hierarchy: an air passenger definition', *Economic Geography, 38,* 1-14

Tabb, W.K. (1977) 'The New York City fiscal crisis' in W.K. Tabb and L. Sawers (eds.), *Marxism and the Metropolis: New Perspectives in Urban Political Economy,* Oxford University Press, New York, pp. 241-66

Tarrant, J.R. (1968) *Retail Distribution in Eastern Yorkshire in Relation to Central Place Theory: A Methodological Approach,* Occasional Papers in Geography 8, University of Hull, Hull

Taylor, E.G.R. (1938) 'Discussion on the geographical distribution of industry', *Geographical Journal, 92,* 18-25

Taylor, T.G. (1949) *Urban Geography: A Study of Site, Evolution, Pattern and Classification in Villages, Towns and Cities,* Methuen, London

Thorngren, B. (1970) 'How do contact systems affect regional development?', *Environment and Planning, 2,* 409-27

Thunen, J.H. von (1826) *Der Isolierte Staat in Berziehung auf Landwirthschaft und Nationalökonomie,* Cotta, Hamburg; translation Wartenburg, C.M., ed. Hall, P. (1966) *Von Thünen's Isolated State,* Pergamon, Oxford

Timms, D.W.G. (1971) *The Urban Mosaic,* Cambridge University Press, Cambridge

Tisdale, H. (1942) 'The process of urbanisation', *Social Forces, 20,* 311-16

Tönnies, F. (1887) *Gemeinschaft and Gesellschaft.* Translated by C.P. Loomis (1955) as *Community and Association,* Routledge & Kegan Paul, London

Tornqvist, G. (1970) *Contact Systems and Regional Development,* Lund Studies in Geography, B, 85, Gleerup, Lund

Ullman, E.L. (1962) *American Commodity Flow: A Geographic Interpretation of Rail and Water Traffic Based upon Principles of Spatial Interchange,* University of Washington Press, Seattle

Ullman, E.L. and Dacey, M.F. (1962) 'The minimum requirements approach to the urban economic base' in K. Norborg (ed.), *Proceedings of the I.G.U.*

Symposium on Urban Geography, C.W.K. Gleerup, Lund, pp. 485-518
United Nations (1955) *Demographic Yearbook, 1952,* United Nations, New York
—— (1974) *Demographic Yearbook, 1974,* United Nations, New York, pp. 9-12
—— (1977) *Demographic Yearbook, 1977,* United Nations, New York
Utton, M.A. (1970) *Industrial Concentration,* Penguin, Harmondsworth
Van Arsdol, M., Camilleri, S.F. and Schmid, C.F. (1958) 'The generality of urban social area indices', *American Sociological Review, 23,* 277-84
Vriser, I. (1971) 'The pattern of central places in Yugoslavia', *Tijschrift voor Economische er Sociale Geografie, 62,* 290-300
Wagon, D.J. and Wilson, A.G. (1971) 'The mathematical model', *Technical Working Paper 5,* SELNEC Transportation Study, Town Hall, Manchester
Warneryd, O. (1968) *Interdependence in Urban Systems,* Regionkonsult Aktiebolog, Gothenburg
Warren, K. (1973) *The American Steel Industry 1850-1970,* Oxford University Press, Oxford
Webber, M.M. (1964) 'The urban place and the non-place urban realm' in M.M. Webber, J.W. Dyckham, D.L. Foley, A.Z. Guttenberg, W.L.C. Wheaton and C.B. Wurster, *Explorations into Urban Structure,* University of Pennsylvania Press, Philadelphia, pp. 79-153
Weber, A. (1909) *Theory of the Location of Industries,* University of Chicago Press, Chicago
Weber A.F. (1899) *The Growth of Cities in the Ninettenth Century,* Macmillan, New York
Weber, M. (1920) *Wirtschaft und Gesellschaft,* Mohr-Siebeck, Tübingen
Weir, S. (1976) 'Red line districts', *Roof,* July, 109-14
Williams, P.R. (1976) 'The role of institutions in the inner London housing market: the case of Islington', *Transactions, Institute of British Geographers, NS 1,* 72-82
—— (1978) 'Building societies and the inner city', *Transactions of the Institute of British Geographers, NS 3,* 23-34
Williams, W.M. (1956) *The Sociology of an English Village: Gosforth,* Routledge & Kegan Paul, London
Wilson, A.G. (1970) *Entropy in Urban and Regional Modelling,* Pion, London
Wilson, A.G., Hawkins, A.F., Hill, G.J. and Wagon, D.J. (1969) 'Calibrating and testing the SELNEC transport model', *Regional Studies, 3,* 337-50. Reprinted in A.G. Wilson (1972), *Papers in Urban and Regional Analysis,* Pion, London, pp. 202-15
Wilson, G. and Wilson, M. (1945) *The Analysis of Social Change,* Cambridge University Press, Cambridge
Wilson, W.H. (1980) 'The ideology, aesthetics and politics of the City Beautiful movement' in A. Sutcliffe (ed.), *The Rise of Modern Urban Planning 1800-1914,* Mansell, London
Wirth, L. (1928) *The Ghetto,* University of Chicago Press, Chicago
—— (1938) 'Urbanism as a way of life', *American Journal of Sociology, 44,* 1-24. Reprinted in P.K. Hatt and A.J. Reiss (eds.) (1957), *Cities and Society,* Free Press, New York
Wolfe, J.M., Driver, G. and Skelton, I. (1980) 'Inner city real estate activity in Montreal: Institutional characteristics of decline', *Canadian Geographer, 4,* 349-67

228   *References*

Wolpert, J. (1964) 'The decision process in a spatial context', *Annals, Association of American Geographers, 54,* 337-58

Wootton, H.J. and Pick, G.W. (1967) 'A model for trip generated by households', *Journal of Transport Economics and Policy, 1,* 137-53

Young, M. and Willmott, P. (1957) *Family and Kinship in East London,* Routledge & Kegan Paul, London

Zipf, G.K. (1949) *Human Behavior and the Principle of Least Effort,* Addison & Wesley, New York

Zorbaugh, H.W. (1929) *The Gold Coast and the Slum,* University of Chicago Press, Chicago

INDEX